T0181741

Studies in Computational Intelligence

Volume 712

Series editor

Janusz Kacprzyk, Polish Academy of Sciences, Warsaw, Poland
e-mail: kacprzyk@ibspan.waw.pl

About this Series

The series "Studies in Computational Intelligence" (SCI) publishes new developments and advances in the various areas of computational intelligence—quickly and with a high quality. The intent is to cover the theory, applications, and design methods of computational intelligence, as embedded in the fields of engineering, computer science, physics and life sciences, as well as the methodologies behind them. The series contains monographs, lecture notes and edited volumes in computational intelligence spanning the areas of neural networks, connectionist systems, genetic algorithms, evolutionary computation, artificial intelligence, cellular automata, self-organizing systems, soft computing, fuzzy systems, and hybrid intelligent systems. Of particular value to both the contributors and the readership are the short publication timeframe and the worldwide distribution, which enable both wide and rapid dissemination of research output.

More information about this series at http://www.springer.com/series/7092

Sankar K. Pal · Shubhra S. Ray
Avatharam Ganivada

Granular Neural Networks, Pattern Recognition and Bioinformatics

 Springer

Sankar K. Pal
Center for Soft Computing Research
Indian Statistical Institute
Kolkata
India

Avatharam Ganivada
Center for Soft Computing Research
Indian Statistical Institute
Kolkata
India

Shubhra S. Ray
Center for Soft Computing Research
Indian Statistical Institute
Kolkata
India

ISSN 1860-949X   ISSN 1860-9503   (electronic)
Studies in Computational Intelligence
ISBN 978-3-319-86079-4   ISBN 978-3-319-57115-7   (eBook)
DOI 10.1007/978-3-319-57115-7

This Springer imprint is published by Springer Nature
The registered company is Springer International Publishing AG
The registered company address is: Gewerbestrasse 11, 6330 Cham, Switzerland

To our parents

Preface

The volume "Granular Neural Networks, Pattern Recognition and Bioinformatics" is an outcome of the granular computing research initiated in 2005 at the Center for Soft Computing Research: A National Facility, Indian Statistical Institute (ISI), Kolkata. The center was established in 2005 by the Department of Science and Technology, Govt. of India under its prestigious IRHPA (Intensification of Research in High Priority Area) program. Now it is an Affiliated Institute of ISI.

Granulation is a process like self-production, self-organization, functioning of brain, Darwinian evolution, group behavior and morphogenesis—which are abstracted from natural phenomena. Accordingly, it has become a component of natural computing. Granulation is inherent in human thinking and reasoning process, and plays an essential role in human cognition. Granular computing (GrC) is a problem-solving paradigm dealing with the basic elements, called granules. A granule may be defined as the clump of indistinguishable elements that are drawn together, for example, by indiscernibility, similarly, proximity or functionality. Granules with different levels of granularity, as determined by its size and shape, may represent a system differently. Since in GrC, computations are performed on granules, rather than on individual data points, computation time is greatly reduced. This made GrC a very useful framework for designing scalable pattern recognition and data mining algorithms for handling large data sets.

The theory of rough sets that deals with a set (concept) defined over a granulated domain provides an effective tool for extracting knowledge from databases. Two of the important characteristics of this theory that drew the attention of researchers in pattern recognition and decision science are its capability of uncertainty handling and granular computing. While the concept of granular computing is inherent in this theory where the granules are defined by equivalence relations, uncertainty arising from the indiscernibility in the universe of discourse can be handled using the concept of lower and upper approximations of the set. Lower and upper approximate regions respectively denote the granules which definitely, and definitely and possibly belong to the set. In real-life problems the set and granules, either or both, could be fuzzy; thereby resulting in fuzzy-lower and fuzzy-upper approximate regions, characterized by membership functions.

Granular neural networks described in the present book are pivoted on the characteristics of lower approximate regions of classes demonstrating its significance. The basic principle of design is—detect lower approximations of classes (regions where the class belonging of samples is certain); find class information granules, called knowledge; form basic networks based on those information, i.e., by knowledge encoding; and then grow the network with samples belonging to upper approximate regions (i.e., samples of possible as well as definite belonging). Information granules considered are fuzzy to deal with real-life problems. The class boundaries generated in this way provide optimum error rate. The networks thus developed are capable of efficient and speedy learning with enhanced performance. These systems have a strong promise to Big data analysis.

The volume, consisting of seven chapters, provides a treatise in a unified framework in this regard, and describes how fuzzy rough granular neural network technologies can be judiciously formulated and used in building efficient pattern recognition and mining models. Formation of granules in the notion of both fuzzy and rough sets is stated. Judicious integration in forming fuzzy-rough information granules based on lower approximate regions enables the network in determining the exactness in class shape as well as handling the uncertainties arising from overlapping regions. Layered network and self-organizing map are considered as basic networks.

Based on the existing as well as new results, the book is structured according to the major phases of a pattern recognition system (e.g., classification, clustering, and feature selection) with a balanced mixture of theory, algorithm and application. Chapter 1 introduces granular computing, pattern recognition and data mining for the convenience of readers. Beginning with the concept of natural computing, the chapter describes in detail the various characteristics and facets of granular computing, granular information processing aspects of natural computing, its different components such as fuzzy sets, rough sets and artificial networks, relevance of granular neural networks, different integrated granular information processing systems, and finally the basic components of pattern recognition and data mining, and big data issues. Chapter 2 deals with classification task, Chaps. 3 and 5 address clustering problems, and Chap. 4 describes feature selection methodologies, all from the point of designing fuzzy rough granular neural network models. Special emphasis has been given to dealing with problems in bioinformatics, e.g., gene analysis and RNA secondary structure prediction, with a possible use of the granular computing paradigm. These are described in Chaps. 6 and 7 respectively. New indices for cluster evaluation and gene ranking are defined. Extensive experimental results have been provided to demonstrate the salient characteristics of the models.

Most of the texts presented in this book are from our published research work. The related and relevant existing approaches or techniques are included wherever necessary. Directions for future research in the concerned topic are provided. A comprehensive bibliography on the subject is appended in each chapter, for the convenience of readers. References to some of the studies in the related areas might have been omitted because of oversight or ignorance.

The book, which is unique in its character, will be useful to graduate students and researchers in computer science, electrical engineering, system science, data science, medical science, bioinformatics and information technology both as a textbook and a reference book for some parts of the curriculum. The researchers and practitioners in industry and R&D laboratories working in the fields of system design, pattern recognition, big data analytics, image analysis, data mining, social network analysis, computational biology, and soft computing or computational intelligence will also be benefited.

Thanks to the co-authors, Dr. Avatharam Ganivada for generating various new ideas in designing granular network models and Dr. Shubhra S. Ray for his valuable contributions to bioinformatics. It is the untiring hard work and dedication of Avatharam during the last ten days that made it possible to complete the manuscript and submit to Springer in time.

We take this opportunity to acknowledge the appreciation of Prof. Janusz Kacprzyk in accepting the book to publish under the SCI (Studies in Computational Intelligence) series of Springer, and Prof. Andrzej Skowron, Warsaw University, Poland for his encouragement and support in the endeavour. We owe a vote of thanks to Dr. Thomas Ditzinger and Dr. Lavanya Diaz of Springer for coordinating the project, as well as the office staff of our Soft Computing Research Center for their support. The book was written when Prof. S.K. Pal held J.C. Bose Fellowship and Raja Ramanna Fellowship of the Govt. of India.

Kolkata, India   Sankar K. Pal
January 2017   Principal Investigator
                                                        Center for Soft Computing Research
                                                              Indian Statistical Institute

Contents

About the Authors

Sankar K. Pal is a Distinguished Scientist and former Director of Indian Statistical Institute. He is currently a DAE Raja Ramanna Fellow and J.C. Bose National Fellow. He founded the Machine Intelligence Unit and the Center for Soft Computing Research: A National Facility in the Institute in Calcutta. He received a Ph.D. in Radio Physics and Electronics from the University of Calcutta in 1979, and another Ph.D. in Electrical Engineering along with DIC from Imperial College, University of London in 1982. He joined his Institute in 1975 as a CSIR Senior Research Fellow where he became a Full Professor in 1987, Distinguished Scientist in 1998 and the Director for the term 2005–2010.

He worked at the University of California, Berkeley and the University of Maryland, College Park in 1986–1987; the NASA Johnson Space Center, Houston, Texas in 1990–1992 and 1994; and in US Naval Research Laboratory, Washington DC in 2004. Since 1997 he has been serving as a Distinguished Visitor of IEEE Computer Society (USA) for the Asia-Pacific Region, and held several visiting positions in Italy, Poland, Hong Kong and Australian universities.

Professor Pal is a Life Fellow of the IEEE, and Fellow of the World Academy of Sciences (TWAS), International Association for Pattern recognition, International Association of Fuzzy Systems, International Rough Set Society, and all the four National Academies for Science/Engineering in India. He is a co-author of 20 books and more than

400 research publications in the areas of pattern recognition and machine learning, image processing, data mining and web intelligence, soft computing, neural nets, genetic algorithms, fuzzy sets, rough sets, cognitive machine and bioinformatics. He visited more than 40 countries as a keynote/invited speaker or an academic visitor.

He has received the S.S. Bhatnagar Prize in 1990 (which is the most coveted award for a scientist in India), Padma Shri in 2013 (one of the highest civilian awards) by the President of India and many prestigious awards in India and abroad including the G.D. Birla Award in 1999, Om Bhasin Award in 1998, Jawaharlal Nehru Fellowship in 1993, Khwarizmi International Award from the President of Iran in 2000, FICCI Award in 2000–2001, Vikram Sarabhai Research Award in 1993, NASA Tech Brief Award (USA) in 1993, IEEE Trans. Neural Networks Outstanding Paper Award in 1994, NASA Patent Application Award (USA) in 1995, IETE-R.L. Wadhwa Gold Medal in 1997, INSA-S.H. Zaheer Medal in 2001, 2005–06 Indian Science Congress-P.C. Mahalanobis Birth Centenary Gold Medal from the Prime Minister of India for Lifetime Achievement, 2007 J.C. Bose Fellowship of the Government of India, Indian National Academy of Engineering (INAE) Chair Professorship in 2013, IETE Diamond Jubilee Medal in 2013, IEEE Fellow Class Golden Jubilee Medal in 2014, INAE-S.N. Mitra Award in 2015, and INSA-Jawaharlal Nehru Birth Centenary Lecture Award in 2017.

Professor Pal is/was an Associate Editor of IEEE Trans. Pattern Analysis and Machine Intelligence (2002–2006), IEEE Trans. Neural Networks [1994–1998 and 2003–2006], Neurocomputing (1995–2005), Pattern Recognition Letters (1993–2011), Int. J. Pattern Recognition & Artificial Intelligence, Applied Intelligence, Information Sciences, Fuzzy Sets and Systems, Fundamenta Informaticae, LNCS Trans. Rough Sets, Int. J. Computational Intelligence and Applications, IET Image Processing, Ingeniería y Ciencia, and J. Intelligent Information Systems; Editor-in-Chief, Int. J. Signal Processing, Image Processing and Pattern Recognition; a Book Series Editor, Frontiers in Artificial Intelligence and Applications, IOS Press, and Statistical Science and

Interdisciplinary Research, World Scientific; a Member, Executive Advisory Editorial Board, IEEE Trans. Fuzzy Systems, Int. Journal on Image and Graphics, and Int. Journal of Approximate Reasoning; and a Guest Editor of IEEE Computer, IEEE SMC and Theoretical Computer Science. (http://www.isical.ac.in/~sankar)

Shubhra Sankar Ray is Associate Professor in the Machine Intelligence Unit, Indian Statistical Institute, Kolkata, and is also associated with its Center for Soft Computing Research. He received M.Sc. in Electronic Science and M.Tech. in Radio Physics and Electronics in 2000 and 2002, respectively from University of Calcutta, and Ph.D. (Engg.) in 2008 from Jadavpur University, Calcutta. He worked as a postdoc fellow at Saha Institute of Nuclear Physics, Calcutta, during 2008–2009. His current research interests include bioinformatics, neural networks, genetic algorithms and soft computing. Three of his publications are listed as curated paper in Saccharomyces Genome Database, Stanford University, USA. He is a recipient of the Microsoft Young Faculty Award in 2010.

Avatharam Ganivada received B.Sc. (Hons) in Statistics, and M.Sc. in Mathematics in 2003 and 2005, respectively, from Andhra University, Visakhapatnam, India, M.Tech. in Comp. Sc. & Tech. in 2008 from the University of Mysore, India, and Ph.D. degree in Computer Science & Engineering in 2016 from University of Calcutta, India. He was with the Center for Soft Computing Research, Indian Statistical Institute, Kolkata during 2009–2015 as Visiting Scientist. He is working at present in ProKarma Softtech Pvt. Ltd., Hyderabad, India. His research areas include fuzzy rough sets, neural networks, pattern recognition and bioinformatics.

Chapter 1
Introduction to Granular Computing, Pattern Recognition and Data Mining

1.1 Introduction

Natural Computing is a consortium of different methods and theories that are emerged from natural phenomena such as brain modeling, self-organization, self-repetition, self-evaluation, self-reproduction, group behavior, Darwinian survival, granulation and perception. Based on the methods abstracted from the phenomena, various computing paradigms/technologies like fractal geometry, DNA computing, quantum computing and granular computing are developed. They take inspiration from the nature for their computation, and are used for solving real life problems. For example, granulation abstracted from natural phenomena possesses perception based information representation which is an inherent characteristic of human thinking and reasoning process. One can refer to [46, 52] for using fuzzy information granulation as a way to natural computing.

In recent years, the nature inspired models or methodologies, involving natural computation, for solving problems are developed those provide a connection between computer science and natural science. It is understood that these models are not the alternative methods; rather, they have been proven substantially as a much more efficient paradigm to deal with various complex tasks. The methodologies are perception-based computing, granular computing, artificial neural networks, granular neural networks, evolutionary computing, artificial immune systems and many others. For example, perception-based computing provides the capability to compute and reason with perception-based information as humans do to perform everyday a wide variety of physical and mental tasks without any specific measurement and computation. The tasks include, for example, driving a car in traffic, parking a car, and playing games. Reflecting the finite ability of the sensory organs and (finally the brain) to resolve details, perceptions are inherently fuzzy-granular (f-granular) [94]. That is, the boundaries of perceived classes are unsharp and the values of the attributes they can take are granulated (a clump of indistinguishable points/objects) [45]. In this situation, one may find it convenient to consider granules, instead of

© Springer International Publishing AG 2017
S.K. Pal et al., *Granular Neural Networks, Pattern Recognition and Bioinformatics*, Studies in Computational Intelligence 712, DOI 10.1007/978-3-319-57115-7_1

the distinct elements, for its handling. Accordingly, granular computing became an effective framework for the design and implementation of efficient and intelligent information processing systems for various real life decision-making applications. The said framework may be modeled in soft computing using fuzzy sets, rough sets, artificial neural networks and their integrations, among other theories.

Chapter 1 is organized as follows: Granular computing initiatives are described in Sect. 1.2. Section 1.3 starts with uncertainty analysis in the perspective of granular computing using fuzzy sets, rough sets and their integration. Preliminaries of neural networks like architecture of a neuron (single layer perceptron), multilayer perceptron and self organizing systems, are then provided. Section 1.4 briefly describes the conventional granular neural networks along with those developed in rough fuzzy frameworks. Pattern recognition from the perspective of human beings and machine learning is explained in Sect. 1.5. Here, a brief introduction of pattern recognition tasks such as classification, clustering and feature selection is provided. Some basic concepts of knowledge discovery from real life data using pattern recognition methods, designed under soft computing paradigms, are described in Sect. 1.6. Big data issues and scope of the book are presented in the last two sections of the chapter.

1.2 Granular Computing

Granular computing (GrC) is a multidisciplinary field that involves theories, methodologies, models, techniques and tools designed for solving complex problems. It is a problem solving paradigm with a basic element, called granule. Granules are formulated by abstraction of common properties of objects/patterns in data. The properties may include similarity, equality, reflexivity, indiscernibility and proximity. The construction of granules is a crucial process, as granules with different sizes and shapes are responsible for the success of granular computing based models. Further, the inter and intra relationships among granules play important roles.

One of the realizations behind granular computing is that precision is sometimes expensive and not very meaningful in modeling complex problems. When a problem involves incomplete, uncertain, and vague information, it may be difficult to differentiate distinct elements and one may find it convenient to consider granules for its handling. In the following section, we summarize these concepts and components briefly [52].

1.2.1 Granules

The meaning/significance of a granule in granular computing is very similar to that of the concept of any subset, class, object or cluster of a universe. That is, granules may be viewed as composition of elements that are drawn together from the universe based on their characteristics like similarity, indiscernibility and functionality.

Each of the granules, according to its shape and size, and with a certain level of granularity may reflect a specific aspect of the problem or form a portion of the system domain. Granules with different granular levels represent the system differently. For example, an image can be described with three granules at the first level of granularity characterizing the regions of the image with three basic colors, such as red, green and blue. At this level the information of the image may be categorized in a broader way, like greenery or bluish regions. If we go further into more details with respect to colors then each of these three granules (color regions) can be described with their subsequent divisions. As an example, each of such divisions can characterize objects (granules) in a particular color such as tree, grass, bushes, where combination of these object regions forms the greenery region.

1.2.2 Granulation

Granulation is an inherent characteristic of human thinking and reasoning process performed in every day life. Granulation is the main task underlying the processes in human cognition [93]. It is the mechanism of forming larger objects into smaller and smaller into larger based on the problem in hand. This idea is also described in [93] as, "granulation involves a decomposition of whole into parts. Conversely, organization involves an integration of parts into whole". This concept leads to the fact that granular computing involves two basic operations, such as granulation and organization. Granulation starts from the problem space as a whole, partitions the problem into sub-spaces, and constructs the desired granules; while organization puts individual elements/granules together to form blocks and to build granules at expected levels. The criteria for the granulation process determine the action for granulating big granules into smaller or small into bigger. Further, the concept of partition and covering comes in the granulation process. A partition consists of disjoint subsets of the universe, and a covering consists of possibly overlap subsets. Partitions are a special type of coverings. Operations on partitions and coverings have been investigated in [86].

1.2.3 Granular Relationships

Granular relationship among granules is a key factor in the process of granulation, as one needs to understand it very precisely for better solution. Granular relationship can be broadly categorized into two groups, such as inter-relation and intra-relation. The former is the basis of grouping small objects together to construct a larger granule based on such as similarity, indistinguishability and functionality; while the latter concerns the granulation of a large granule into smaller units and the interactions between components of a granule as well. A granule is a refinement of another granule if it is contained in the latter. Similarly, the latter is called coarsening of the former. These relationships function like set containment in the set based domains.

1.2.4 Computation with Granules

Computation with granules is the final step in granular computing process. Computing and reasoning in various ways with different granules based on their relationship and significance is the basis for the computation method. These operations are broadly categorized as either computations within granules or computations between granules. Computations within granules include finding characterization of granules, e.g. membership functions of fuzzy granules; inducing rules from granules, such as classification rules that characterize the classes of objects; forming concepts that granules entail. On the other hand, computations between granules usually operate on the interrelations between granules, transformations from one granule to another, clustering granules, and dividing granules.

1.3 Granular Information Processing Aspects of Natural Computing

Granular information processing is one of the human-inspired problem solving aspects of natural computing, as information abstraction is inherent in human thinking and reasoning process, and plays an essential role in human cognition [52]. Among the different facets of natural computing fuzzy sets, rough sets, fuzzy rough sets, neural networks and their hybridization are well accepted paradigms that are based on the construction, representation and interpretation of granules. In this section, we provide an overview of these tools, emphasizing the characteristic features and design principles.

1.3.1 Fuzzy Set

The theory of fuzzy sets was introduced by Zadeh [88] to model the uncertainty in natural language. Traditional set theory deals with whether an element "belongs to" or "does not belong to" a set. Fuzzy set theory, on the other hand, concerns with the continuum degree of belonging, and offers a new way to observe and investigate the relation between sets and its members. It is defined as follows:

Let X be a classical set of objects, called the universe. A fuzzy set A in X is a set of ordered pairs $A = \{(x, \mu_A(x)) | x \in X\}$, where $\mu_A : X \rightarrow M$ is called the *membership function* of x in A which maps X to membership space M. Membership $\mu_A(x)$ indicates the degree of similarity (compatibility) of an object x to an imprecise concept, as characterized by the fuzzy set A. The domain of M is $[0, 1]$. If $M = \{0, 1\}$, i.e., the members are only assigned either 0 or 1 membership value, then A possesses the characteristics of a crisp or classical set.

The set of all elements having positive memberships in fuzzy set A constitutes its support set, i.e.,

$$Support(A) = \{x | \mu_A(x) > 0\}. \tag{1.1}$$

The cardinality of the fuzzy set A is defined as

$$|A| = \sum_{x \in X} \mu_A(x). \tag{1.2}$$

Union and intersection of two fuzzy sets A and B are also fuzzy sets and we denote them as $A \cup B$ and $A \cap B$ respectively. The membership functions characterizing the union and intersection of A and B are as follows:

$$\mu_{A \cup B}(x) = \max(\mu_A(x), \mu_B(x)), x \in X \tag{1.3}$$
$$\mu_{A \cap B}(x) = \min(\mu_A(x), \mu_B(x)), x \in X. \tag{1.4}$$

Fuzzy logic is a system of reason and computation in which the objects of reasoning and computation are classes with unsharp borders or classes with no discernable borders i.e., the objects are characterized by fuzzy sets. Humans have a remarkable capability to reason, compute, and make rational decisions based on information which consists of perceptions not measurements or information which is described in natural language. Basically, a natural language is a system for describing perceptions. Traditional logical systems and probability theory do not have this capability. Progression to achievement to this capability is one of the principle objectives of Fuzzy Logic.

In summary,

- Theory of Fuzzy sets (FS) provides a generalization of classical set theory.
- Fuzzy sets are nothing but membership functions.
- Membership functions are context dependent.
- Membership value concerns with the compatibility (similarity) of an element x with an imprecise concept represented by the set, unlike the probability of occurrence of x which concerns with the no. of occurrence of x.
- Flexibility of FS is associated with the concept of membership function. Higher the membership value of an object is, lower is the amount of stretching required to fit the object with the imprecise concept representing the FS.

The concept of information granulation with the advent of fuzzy set theory and the formulation of the methodology for fuzzy information granulation is described in [93]. Fuzzy information granulation deals with the construction of fuzzy granules for a particular level of granulation based on the problem in hand. Although crisp or discrete granules have wide range of applications, the human thinking and reasoning process are better formulated with fuzzy granules. In fact, fuzzy information granulation is the key to fuzzy logic, as it is the key to human thinking and reasoning process. Fuzzy information granulation plays a vital role in the conception and design of intelligent systems. It is important to mention here that there is a long list of tasks which humans can perform with no difficulty, whereas they may be a massive task for machine to perform without the use of fuzzy information granulation [93].

1.3.1.1   Fuzzy Modeling

A fuzzy logic based system can be characterized as the nonlinear mapping of an input data set to a scalar output data. A fuzzy system consists of four main parts: fuzzifier, rules, inference engine, and defuzzifier. In the process, firstly, a crisp set of input data is converted to a fuzzy set using fuzzy linguistic variables, fuzzy linguistic terms and membership functions. This step is known as fuzzification. Next, an inference is made based on a set of rules. Lastly, the resulting fuzzy output is mapped to a crisp output in the defuzzification step.

1.3.1.2   Fuzzy Information Granulation Models

In the recent past, several attempts have been made in the construction of fuzzy granules with desired level of granulation. The process of fuzzy granulation or f-granulation involves the basic idea of generating a family of fuzzy granules from numerical features and transforming them into fuzzy linguistic variables. These variables thus keep the semantics of the data and are easy to understand.

Fuzzy information granulation has come up as an important concept in fuzzy set theories, rough set theories and the combination of both in recent years [40, 60, 61, 93]. In general, the process of fuzzy granulation can be broadly categorized as class-dependent (CD) and class-independent (CI). With CI granulation each feature space is granulated/described irrespective of the classes of available patterns. The granulation process follows the concept of "decomposition of whole into parts", and information of feature is described with the membership values corresponding to the linguistic properties. One can describe the feature with one or more number of overlapping linguistic variables with different properties for desired level of granulation. For example, the input feature is fuzzified in [57] with the overlapping partitions of the linguistic properties low, medium and high. These overlapping functions along each of the axes generate the fuzzy granulated feature space in n-dimension and the granulated space contains $G \times n$ granules, where G is the number of linguistic properties. The degree of belonging of a pattern to a granule is determined by the corresponding membership function. While this model is effective in representing input in any form in a knowledge based system, the process of granulation does not take care of the class belonging information of features to different classes. This may lead to a degradation of performance in a decision making process, particularly for data sets with highly overlapping classes. On the other hand, in CD granulation, each feature explores its class belonging information to different classes. In this process, features are described by the fuzzy sets equal to the number of classes, and the generated fuzzy granules restore the individual class information. These are

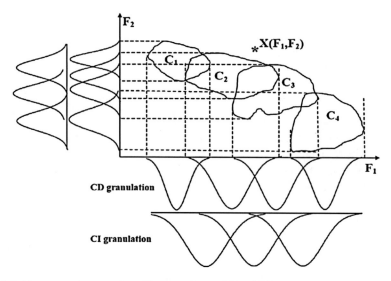

Fig. 1.1 Fuzzy granule generation. The figure is opted from [52]

described in Fig. 1.1 for four classes in a two dimensional plane. Note that the number of granules generated in CD granulation is C^n where C is the number of classes, whereas for CI granulation it is 3^n corresponding to three linguistic properties *low*, *medium* and *high* in an n dimensional plane.

These granulated features, thus generated, can be considered as input to any classifier for classification. An example application of this concept is demonstrated in [53] with various patterns of speech data and remote sensing images. A fuzzy granular computing framework, in which linguistic variables (information granules) are used as initial connection weights of neural network, is provided in [15]. The linguistic variable consists of a 5-tuple (X, U, T, S, M) where X is a fuzzy set, U is a universe, T is a set of atomic terms, S is syntactic rule and M is a semantic rule. Here an atomic term represents a word or union of a set of words [90]. The syntactic rule S is used to generate new atomic terms based on linguistic arithmetic operations performed between any two atomic terms T.

1.3.1.3   Applications

Over the years, fuzzy logic has been applied successfully in many areas including

- Pattern recognition and classification
- Image and speech processing
- Fuzzy systems for prediction
- Fuzzy control
- Monitoring

- Diagnosis
- Optimization and decision making
- Group decision making
- Hybrid intelligence system

The application areas can be broadly grouped in three categories such as fuzzy logic control, fuzzy expert systems, and fuzzy pattern recognition and image processing. In fuzzy control systems, the aim is to replace/emulate human operators, in fuzzy expert system design the aim is to replace/emulate human expertise, whereas the fuzzy pattern recognition and image processing systems incorporate generalization of crisp decisions and uncertainty handling.

One may note that the first paper of Zadeh in pattern recognition appeared in 1966 [5] following a technical report in 1965, whereas his first paper in fuzzy control appeared in 1973 [89]. It means Zadeh had the concept (notion) of fuzzy classification mainly when he was developing the theory, although his fuzzy control systems got massive success in 1980's because of its implementation in Japanese products. Since techniques of pattern recognition and image processing interact with and support a large percentage of control systems (e.g., Mars rover control, camera tracking and docking at space shuttles, fuzzy camcorders, traffic flow control), applications of pattern recognition and vision systems have matured, especially because of the commercial success of Japanese products based on fuzzy logic control.

1.3.2 Rough Set

Let X be a classical set of objects, in a universe of discourse U. Under situations when relations exist among elements of U, X might not be exactly definable in U as some elements of U that belong to the set X might be related to some other elements of U that do not belong to set X.

When a relation, say R, exists among elements of U, limited discernibility draws elements of U together governed by the relation R resulting in the formulation of granules in U. Here, a set of elements in U that are indiscernible from or related to each other is referred to as a granule. Let us represent granules using Y and the family of all granules formed due to the relation R using U/R.

As mentioned earlier, the relation R among elements of U might result in an inexact definition of X. To tackle such cases in the framework of rough set theory of Pawlak [64], X is approximately represented by two exactly definable sets $\underline{R}X$ and $\overline{R}X$ in U given as !equivalance classes!example of rough set

$$\underline{R}X = \bigcup \{Y \in U/R | Y \subseteq X\} \tag{1.5}$$

$$\overline{R}X = \bigcup \{Y \in U/R | Y \cap X \neq \}. \tag{1.6}$$

In the above, the set $\underline{R}X$ is defined by the union of all granules that are subsets of the set X and the set $\overline{R}X$ is defined by the union of all granules that have non-empty intersection with the set X. The sets $\underline{R}X$ and $\overline{R}X$ are respectively called the *lower approximation* and *upper approximation* of X with the imprecise concept R.

In rough set theory [64], an information system is considered as an attribute value table where rows indicate objects of the universe and columns denotes attributes. It is denoted by $(U, \mathcal{A})$ where universe U is a nonempty finite set of objects, and $\mathcal{A}$ is a non empty finite set of conditional attributes. For every $\mathcal{B} \subseteq \mathcal{A}$, the $\mathcal{B}$-indiscernibility relation, $R_\mathcal{B}$, is defined as

$$R_\mathcal{B} = \{(x, y) \in U^2 \mid \forall a \in \mathcal{B}, \quad a(x) = a(y)\}, \tag{1.7}$$

This implies that, the patterns x and $y \in U \times U$ are called indiscernible with respect to the attribute $a \in \mathcal{B}$, when the value of x is equal to that of y with respect to the attribute a. Based on the indiscernibility relation, the crisp equivalence classes/tolerance relations, denoted by $[\tilde{x}]_{R_\mathcal{B}}$, are generated corresponding to an attribute $\mathcal{B}$. The equivalence classes can be regarded as equivalence granule. The granule contains a set of objects whose values are equal corresponding to an attribute a. Given $A \subseteq U$ an arbitrary set, it may not able to describe U directly using the equivalence classes. In this case, one may characterize A by a pair of lower and upper approximations, i.e.,

$$R_\mathcal{B} \downarrow A = \{x \in U | [\tilde{x}]_{R_\mathcal{B}} \subseteq A\}, \; and \tag{1.8}$$

$$R_\mathcal{B} \uparrow A = \{x \in U | [\tilde{x}]_{R_\mathcal{B}} \cap A \neq \emptyset\}. \tag{1.9}$$

The pair $(R_\mathcal{B} \downarrow A, R_\mathcal{B} \uparrow A)$ is called a rough set of A, given the indiscernibility relation $R_\mathcal{B}$. In rough set theory, a decision system is characterized by $(U, \mathcal{A} \cup d)$ where d $(d \notin \mathcal{A})$ is called a decision attribute and its equivalence classes $[\tilde{x}]_{R_d}$ are called decision classes (decision granules).

It may be mentioned here that:

- Fuzzy set and rough set are reputed to handle uncertainties arising from two different aspects, namely, overlapping concepts (or characters) and granularity in the domain of discourse respectively. While the former uses the notion of class membership of an element, the latter revolves around the concept of approximating from lower and upper sides of a set defined over a granular domain.
- One may cite hundreds of readily examples for fuzzy set, e.g., large number, long street, beautiful lady, sharp curve and young man. Whereas, this is not so true for rough set. Though the name of the set is rough [64], basically it is a crisp set with rough descriptions.
- *Example of a rough set*: Let us consider a school consisting of a number of sections with different strengths of students. Here each section can be viewed as a granule where its constituting students are selected by some similarity relation in terms of their age or sex or marks obtained in the previous class examination.

Suppose, a contaminated disease has spread over the school. The principal of the school made a decision that the "infected classes should be declared holiday". This lead to two possibilities, namely,

- if all the students in a section are affected. Let A denote the set of all such sections.
- if at least one of the students in a section is affected. Let B denote the set of all such sections. Obviously, B would include A.

Here A and B denote the lower and upper approximations of the crisp concept (set)—"declare holiday for infected classes". Students in A have definitely to be declared holiday, with no question. Doubt may arise only for those students belonging to the set $B - A$ where at least one student, but not all, is affected.

The main advantage of rough set based granulation in data analysis is that it enables the discovery of data dependencies and performs the reduction/selection of features contained in a data set using the data alone, requiring no additional information, such as basic probability assignment in Dempster-Shafer theory, grade of membership or the value of possibility in fuzzy set theory. In addition, the rough set based granular method has many important advantages, such as it finds hidden patterns in data, finds minimal sets of data (data reduction), evaluates the significance of data, generates sets of decision rules from data and facilitates the interpretation of obtained results.

1.3.2.1   Rough Mereology

Rough mereological granules are constructed from rough inclusions and to approximate partial containment operators [70, 71]. Rough mereology (RM) is a method for synthesis and analysis of objects in the distributed environments of intelligent agents. In a complex environment like synthesis of objects for a given specification to a satisfactory degree, or for control, rough mereology plays a key role. Moreover, it has been used recently for developing foundations of the information granule calculus. RM is an attempt towards formalization of the paradigm of computing with words based on perception, as formulated by Zadeh [92, 93].

RM is built based on the inclusion relation. This method is also considered as approximate calculus of parts, an approach to reasoning under uncertainty based on the notion of an approximate part (part to a degree). The paradigms of granular computing, computing with words and spatial reasoning are particularly suited to a unified treatment by means of RM. It is a generalization of the rough set and fuzzy set approaches [70, 71]. This relation can be used to define other basic concepts like closeness of information granules, their semantics, indiscernibility/discernibility of objects, information granule approximation and approximation spaces, perception structure of information granules as well as the notion of ontology approximation. The rough inclusion relations together with operations for construction of new information granules from already existing ones create a core of a calculus of information granules.

1.3.2.2 Rough Modeling

The basic assumption of rough set theory is that human knowledge about a universe depends upon their capability to classify its objects, and classifications of a universe and equivalence relations defined on it are known to be interchangeable notions. To improve the modeling capability of basic rough set theory, several extensions have been made in different directions. These extensions are as follows:

The granulation of objects induced by an equivalent relation is a set of equivalence classes, in which each equivalence class can be regarded as an information granule, as described in [64]. One extension to this is based on tolerance relations instead of equivalence relations. These rough sets are sometime called incomplete rough sets. The granulation of objects induced by a tolerance relation generates a set of tolerance classes, in which each tolerance class also can be seen as a tolerance information granule. Another method of information granulation is characterized by using a general binary relation, where objects are granulated into a set of information granules, called a binary granular structure.

With the notion of granular computing (GrC), a general concept portrayed by Pawlak's rough set is always characterized with upper and lower approximations under a single granulation. The basic operation involved in rough set is that it partitions the object space based on a feature set using some equivalence relation. The partition spaces thus generated are also known as granules, which become the elemental building blocks for data analysis. The rough representation of a set with upper and lower approximations is shown in Fig. 1.2. Thus the concept is depicted by known knowledge induced by a single relation on the universe, which includes equivalence relation, tolerance relation, reflexive relation, and many more. This clearly states that given any such relations, one can determine a certain granular structure (or called a granular space). Pawlaks rough set generally takes the following assumption in describing an objective/decision/target concept.

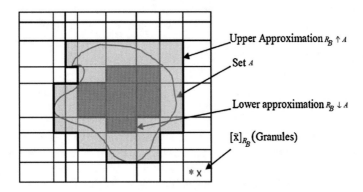

Fig. 1.2 Rough representation of a set with upper and lower approximations

Let $\mathcal{A}$ and $\mathcal{B}$ be two conditional sets of features and $A \subseteq U$ be a decision feature, then the rough set of A is derived from the quotient set $U/(\mathcal{A} \cup \mathcal{B})$. The quotient set is expressed as

$$\widehat{\mathcal{A} \cup \mathcal{B}} = A_i \bigcap B_i : A_i \in U/\mathcal{A}, B_i \in U/\mathcal{B}, A_i \bigcap B_i \neq \emptyset \qquad (1.10)$$

This relation clearly implies that an intersection operation can be performed between any A_i and B_i, and the decision feature is approximately described by using the quotient set $U/(\mathcal{A} \cup \mathcal{B})$. Moreover, the decision feature can be described using a partition of the space that generates fine granules through combining two known granulations (partitions) induced from two-attribute subsets. Although it generates a much finer granulation and more knowledge, the combination/fining destroys the original granulation structure/partitions. However, this assumption cannot always be required in practice. Three practical cases are mentioned below to demonstrate its restrictions.

1. CASE 1: For the same object of a data set, if any contradiction or inconsistent relationship exists between its values under two attributes sets $\mathcal{A}$ and $\mathcal{B}$, the intersection operations between their quotient sets and the target concept cannot be approximated by using $U/(\mathcal{A} \cup \mathcal{B})$.
2. CASE 2: For the same object or element, the decisions are different. Under such circumstance, the intersection operations between any two quotient sets will be redundant for decision making.
3. CASE 3: For the reduction of the time complexity of knowledge discovery, it is unnecessary to perform the intersection operations in between all the sites in the context of distributive information systems.

In such cases, the decision features need to be described through multi binary relations (e.g., equivalence relation, tolerance relation, reflexive relation and neighborhood relation) on the universe, and this is purely according to the problem in hand. For many practical issues, rough set theory is applied widely with the concept of multigranulation rough set framework based on multi equivalence relations. For detail description of multigranulation method of approximating the solution, one may refer to [72, 85].

1.3.2.3 Applications

Rough set theory allows characterization of a set of objects in terms of attribute values; finding dependencies (total or partial) between attributes; reduction of superfluous attributes; finding significant attributes; and generation of decision rules. Two of its main characteristics, namely, uncertainty analysis through lower and upper approximations, and granular computing through information granules have drawn the attention of applied scientists. Accordingly, these properties made the theory an attractive choice in dealing with real-time complex problems, such as in:

- Pattern recognition and image processing
- Artificial intelligence
- Data mining and knowledge discovery
- Acoustical analysis
- Power system security analysis
- Spatial and meteorological pattern classification
- Intelligent control systems

1.3.2.4   Rough Information Granulation Models

Several researchers have examined the process of developing information granules using rough set theory. The theory is used in [55] to obtain dependency rules which model various informative regions in the fuzzy granulated feature space. The fuzzy membership function corresponding to these informative regions constitute what are called rough information granules. These granules involve a reduced number of relevant features; thereby resulting in dimensionality reduction. For example, given the circular object region in Fig. 1.3, rough set theory can, whether supervised or unsupervised mode, extract the rule $F_{1M} \bigwedge F_{2M}$ (i.e., feature F_1 is M AND feature F_2 is M) to encode the object region. This rule, which represents the rectangle (shown by bold line), provides a crude description of the object or region and can be viewed as its information granule. Though both the features F_1 and F_2 have appeared here in the rule, in practical scenario with large feature problem, the rough rules corresponding to various regions may not contain all the features. Its application in information retrieval for mining large data sets has been adequately demonstrated [55]. In [72], the concepts of rough sets are also employed for generating multi granulations, based on multi equivalence relations. A neighborhood type of information granule is developed in [53]. Here, a neighborhood rough set defines the granule and its size is controlled by a distance function. Information granulation using covering based rough set is provided in [97].

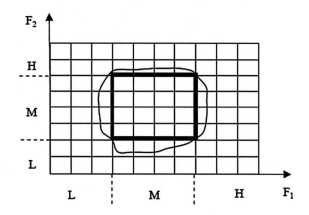

Fig. 1.3 Rough set theoretic rules for an object. L, M and H denote respectively low, medium and high feature values

1.3.3 Fuzzy Rough Sets

Several researchers have explored various relationships to extend and integrate the concepts of fuzzy and rough sets [9, 11, 80]. The judicious integration deals with the additional aspects of data imperfection, such as uncertainty, vagueness and incompleteness especially in the context of granular computing. These may be described with the existing activities being performed every day in human life. For example,

- Uncertainty: The scientists predict many people will die from earthquake expected anytime, today. This would be uncertainty, because (i) many people may not die (ii) there might be no earthquake at all today.
- Vagueness: The government say the earthquake, hit today, killed more than ten thousand people or around twenty thousand people. Here, the words "more than and around" are indicating the vagueness because there is lack of precision with the specified numbers.
- Incompleteness: The scientists claim that there is fifty percent of chance of another quake occurring today. This would be incompleteness, because fifty percent of chance likely to be the incomplete information.

The main purpose of fuzzy rough hybridization/integration is to provide flexible processing capabilities [9, 62], robust solutions and advanced tools for data analysis [38] and a framework for efficient uncertainty handling [4]. Some typical applications of such hybridization for the problem of feature/attribute selection are reported in [11, 32]. In doing so, the following two principles were employed in generalizing the lower and upper approximations in Eqs. 1.8 and 1.9, respectively.

- The set A may be generalized to a fuzzy set in U where the objects belong to a given concept (i.e., subset of the universe) with membership degrees in [0, 1].
- Usually, "objects indistinguishability" (for instance, with respect to their attribute values in an information system) is described by means of an equivalence relation R in U in Pawlak's rough approximation [64]. Rather than doing so, the approximation equality of objects can be represented by a fuzzy similarity relation R in generalized approximation space. As a result, objects are partitioned into different classes or granules with "soft" boundaries on the basis of similarity between them.

As the crisp equivalence classes are central to the rough sets, the fuzzy equivalence classes which are based on fuzzy similarity relation R, are important for fuzzy rough sets. The fuzzy equivalence class represents a fuzzy granule. The decision classes can be defined as crisp or fuzzy, corresponding to decision attribute. Here, the significance of the fuzzy reflexive relation is that it measures the similarity between any two patterns in a universe. There is also a possibility of outliers to be present in the data due to noise. Fuzzy rough sets, based on a fuzzy tolerance relation, provide a means by which discrete or real-valued noisy data can be effectively reduced without any additional information about the data (such as thresholds on a particular domain of universe) for its analysis. The granulation structure produced by an equivalence class provides a partition of the universe. The intension of it is to approximate an imprecise concept in the domain of universe by a pair of approximation concepts, called lower and upper approximations. These approximations are used to define the notion of positive degree of each object and the dependency factor of each conditional attribute, all of which may then be used to extract the domain knowledge about the data.

The formation of the lower and upper approximations of a set involves the fuzzy equivalence classes and decision classes. A notion of fuzzy rough set, denoted by $((R_B \downarrow A)(x), (R_B \uparrow A)(x))$, is typically defined as [16]

$$(R_B \downarrow R_A)(x) = \inf_{y \in U} \max\{1 - R_B(x, y), R_A(x)\} \text{ and} \tag{1.11}$$

$$(R_B \uparrow R_A)(x) = \sup_{y \in U} \min\{R_B(x, y), R_A(x)\} \tag{1.12}$$

The fuzzy similarity relation R, computed between all possible pairs of patterns, defines a similarity matrix corresponding to an attribute where each row in the matrix refers to a fuzzy equivalence granule. The $R_B(x, y)$ is a fuzzy similarity value between the patterns x and y corresponding to an attribute in a set B and $R_A(x)$ is fuzzy decision value of a pattern x corresponding to a decision class (set) A. This implies, a set $A \subset U$ is approximated by assigning membership value to every pattern of the set belonging to its lower and upper approximations, based on the fuzzy similarity relation.

There may be situations where the relation R is crisp, but the set A is fuzzy; and both R and A are fuzzy. Then the corresponding fuzzy lower and upper approximations together constitute what may be called rough fuzzy set and fuzzy rough fuzzy set [80]. Other forms of fuzzy rough set, based on a pair of logical operators, can be found in [43].

1.3.4 Artificial Neural Networks

Human intelligence and discriminating power is mainly attributed to the massively connected network of biological neurons in the human brain. Artificial neural network (ANN) is a system composed of many simple processing elements (nodes) operating in parallel whose function is determined by network structure, connection strength of links (weight), and processing performed at computing elements or nodes. Artificial neural networks (ANNs) attempt to emulate information representation/processing scheme and discrimination ability of neurons (cells) in the human brain. The human brain consists of about 10^{11} neurons containing approximately 60 trillion connections among them [27]. The synapses of connections mediate information (nervous impulse) from one neuron to another neuron. In a similar way, ANNs try to emulate the biological neurons and are composed of processing units which are interconnected with each other. The weights, referred as synapses, are assigned to links connecting the neurons and communicate information between the nodes. The characteristics of neural networks include non linearity, input output mapping, adaptivity, fault tolerance, robustness, ruggedness, speed via massive parallelism and optimality. Some of them are discussed as follows:

Non linearity: The ANNs are capable of handling the non linearly separable data by defining curves.

Input output mapping: During training, the networks provide a mapping from every input sample to an output value which is encoded into the connection weights so as to obtain modified weights.

Adaptivity: The ANNs are flexible to adapt changes in the connection weights between neurons, during training with different values of input, in various operating system environments.

Fault tolerance: The ANNs have the potential to tolerate the faults. The faults mainly concern unfavorable system conditions and networks redundancy. The unfavorable system conditions are insufficient system memory, i.e., not enough memory to execute a program and failure of some neurons/components. The redundancy in neural networks is caused by large number of inter connections between nodes.

Speed: As the nodes of ANNs function in parallel and are composed in feed forward fashion, the networks can achieve high computational speed.

1.3.4.1 Architecture of a Neuron

The architecture of a neuron, l, is shown in Fig. 1.4. It is apparent form the figure that there are three basic components, like, connection weights, an adder function and an activation function. Let $\bar{x} = \{x_1, x_2, \ldots, x_n\}$ denote an input vector, and $\{w_{l1}, w_{l2}, \ldots, w_{ln}\}$ be a set of synaptic weights, connecting links of the neuron l where n is the number of features. The adder function adds the weighted product of the input vector and the weights of the neuron l, denoted by u_l. It is defined as

$$u_l = \sum_{j=1}^{n} w_{lj}x_j + b_0, \tag{1.13}$$

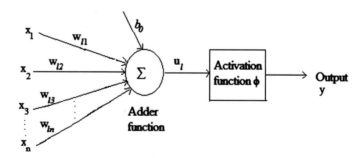

Fig. 1.4 Architecture of a neuron

where the bias b_0 is termed as a synaptic weight connected to the neuron. The actual output corresponding to an input pattern is computed as

$$y = \Phi(u_l) \tag{1.14}$$

where Φ is a non linear activation function and it limits the resulting value of adder function.

Learning procedure:

The learning of the neuron is based on modification of weights connected to the neuron. The weights are modified by training the neuron using a set of input patterns and the error corresponding to the desired responses. The training is performed for different number of iterations, denoted by e, and is explained in the following basic steps.

1. Initialize the synaptic weights w_{lj} of neuron l, shown in the aforesaid figure, with the values lying within [0, 1], for $l = 1, j = 1, 2 \dots, n$. Here n is the no. of features
2. Find the sum of product of input vector and connection weights at the neuron using Eq. 1.13.
3. Find the actual output at the neuron using 1.14.
4. Modify the connection weights of the neuron using

$$w_j(e + 1) = w_j(e) + \eta(d(e) - y(e))x_j, \tag{1.15}$$

$$b_j(e + 1) = b_j(e) + \eta(d(e) - y(e)); \tag{1.16}$$

where e is the number of iterations, the value of the desired output $d(e)$ is, typically, $+1$ when the input pattern/sample belongs to its actual class, and -1 when the input pattern belongs to the other class (other than the actual class). The value of the parameter η is chosen within 0 to 1.
5. Repeat Steps 1–4 until the difference between the desired response (d) and the actual output (y) is minimum.

The modified weights among the nodes, representing the knowledge about training data, are utilized to solve the complex problems.

The other network architectures like, single-layer feedforward networks, multilayer feedforward networks, recurrent networks and lattice structures have been described in [27]. Descriptions of multilayer feedforward networks and self-organizing systems are briefly provided in the following two sections.

1.3.4.2 Multilayer Feedforward Networks

Multilayer feedforward networks consist of input layer, output layer and one or more hidden layers placed between input layer and output layer. A non linear activation function, typically a sigmoidal function, is set to each of the nodes in the hidden and output layers. All the nodes of the network are connected in feed forward fashion.

Fig. 1.5 Architecture of a
three layer neural network

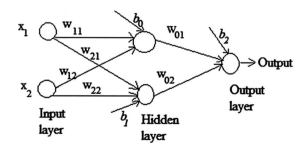

The synaptic weights are used to establish links between input nodes and hidden nodes, and hidden nodes and output nodes.

An architecture of three layered network for classification is shown in Fig. 1.5. It is clear from the figure that the network consists of two input nodes, corresponding to an input vector (feature vector) $\bar{x} = \{x_1, x_2\}$, two hidden nodes and one output node. The connection weights w_{11}, w_{12}, w_{21} and w_{22}, and w_{01} and w_{02} are initialized between the two input nodes and two hidden nodes and two hidden nodes and one output node, respectively, as shown in Fig. 1.5. The biases b_0, b_1 and b_2 are initialized at the first hidden node, second hidden node and output node respectively. It may noted that the hidden layers can be more than one. The aim of this network is to generalize the position (inside or outside the region defined by hyperplanes corresponding to different nodes) of a vector $\bar{x}$. The output of the each input node is a straight line (linear product of input vector and its associated connection weight). The hidden nodes and output node provide hyperplanes. These are derived from the actual outputs of the nodes in the hidden layer and the output layer, based on activation function, Φ. As there are two hidden nodes and one output node in the network, three hyperplanes are generated. A region corresponding to every input pattern is constituted by combing three hyperplanes (decision lines) which are characterized as follows:

$$y_1 = \Phi(x_1 w_{11} + x_2 w_{12} + b_0). \tag{1.17}$$

$$y_2 = \Phi(x_2 w_{21} + x_2 w_{22} + b_1). \tag{1.18}$$

$$y_3 = \Phi(y_1 w_{01} + y_2 w_{02} + b_2). \tag{1.19}$$

The hyperplanes in Eqs. 1.17, 1.18, and 1.19 together constitute a region in which the position of input vector $\bar{x}$ is identified for classification purpose. Further, for easy understanding of the classification process of two non linear separable classes using XOR data, one can see [27] which is based on the network shown in Fig. 1.5.

Fig. 1.6 Architecture of self-organizing map

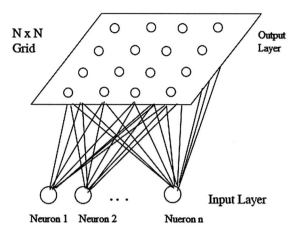

N x N
Grid

Output
Layer

Input Layer

Neuron 1   Neuron 2   Nueron n

1.3.4.3 Self-organizing Systems

The self organizing systems are commonly defined using learning algorithms like Hebbian learning, Kohonen competitive learning and Shanon information theory. The aim of the learning algorithms is to refine the connection weights of the networks and thereby making the networks adaptable for different types of problems. These systems are used for unsupervised classification (clustering). Self organizing map (SOM) [34] is one of the self-organizing systems and is based on competitive learning. The initial architecture of self-organizing map is shown in Fig. 1.6. The SOM is an useful technique for analyzing and visualizing data. It provides a mapping from the higher dimensional feature space to two dimensional space in defining the input layer and the output layer. The nodes in the input layer and output layer are connected in a feed forward fashion. The training involves four basic steps, initialization of connection weights, sampling of input data, similarity matching and updating the connection weights based on competitive learning. During training, the connection weights are initialized with small random numbers within 0 to 1. The input vectors are randomly selected and we present them at the nodes in the input layer. The distances (similarity) between input vector and weight vectors of all the nodes in the output layer are computed. The node which has minimum distance is considered as a winning node among all the output nodes. The neighborhood neurons around the winning neuron are defined and which are updated using Gaussian neighborhood function. After the training of SOM is completed, each node in the output layer is associated with the group of patterns, representing a cluster.

1.3.4.4 Applications

ANN deals with a multi disciplinary field and as such its applications are numerous including the following disciplines:

- Finance
- Industry
- Agriculture
- Physics
- Statistics
- Cognitive science
- Neuroscience
- Weather forecasting
- Computer science and engineering
- Spatial analysis and geography

1.3.4.5 What is Granular Neural Network?

The ANNs are usually trained with a given set of numerical input output data, where the numerical weights of the network are adapted and provide a knowledge mapping between input and output data. The trained network then uses the discovered knowledge to predict the behavior of the output for new sets of input data. ANNs are therefore a useful tool for knowledge discovery mainly because of these characteristics. Quite often, data sets are linguistic or symbolic in nature instead of discrete numerical (as mentioned above) and become unfit for the conventional neural networks. For those cases, the network must be capable of processing both numerical and linguistic data, which turns out to be a challenging task. To deal with this, a new version of neural network called granular neural network (GNN) [61, 67, 96] has been evolved by incorporating the granular concept in the neural network design. In general, the GNN is capable of processing granular data (obtained such as from numerical and linguistic data), extracting granular information, fusing granular data sets, compressing a granular data base, and predicting new information.

In granular data, a granule can be, for example, a class of numbers, a cluster of images, a class of regions, a set of concepts, a group of objects, and a category of data. These granules are input and output of GNN, just like any natural data are input and output of biological neural networks in the human brain. Therefore, GNN seems to be more useful and effective to process natural information of granules than conventional numerical-data-based neural networks.

The basic biological motivations of GNN are

- It enjoys the similarity of ANNs to biological neural networks in many senses and builds models to understand various nerves and brain operations by simulation.
- It mimics various cognitive capabilities of human beings and solves problems, through learning/adaptation, prediction, and optimization.
- It allows for parallel information processing through its massively parallel architecture, and makes the model more robust to fault tolerant.

In a nutshell, the working principle of GNN is as follows: It is the combination of neurons that construct, design and operate at the level of information granules. Development of GNNs involves two main phases: Granulation of numeric data where a collection of information granules is formed, and the construction of the neural network, where any learning that takes place with the neural network should be based on the information granules rather than the original data. However, not much attention has been paid to develop formal frameworks of information granulation applied to the design of neural networks.

In the following section various aspects and versions of GNN are described with different design architectures.

1.4   Integrated Granular Information Processing Systems

Designing an integrated granular information processing system with more than one granular method has been popularly undertaken by the researchers to solve complex problems. The motivation is to combine the merits of individual techniques in order to design a system that can exploit the tolerance for imprecision, uncertainty, approximate reasoning, and partial truth in order to achieve tractability, robustness, and low cost solution in real-life ambiguous situations. Many attempts have been made in this regard, where the merits of the basic methods (neural networks, roughs sets, fuzzy sets and fuzzy rough sets) of information granulation are hybridized in different combinations [61, 62]. During integration each paradigm helps other, rather than competing with. In the following sections, we briefly describe some such integrations with applications.

1.4.1   Fuzzy Granular Neural Network Models

Integration of neural networks (NNs) and fuzzy logic provides a hybrid paradigm known as neuro-fuzzy (NF) computing [58] which is the most visible one realized so far among all other integrations in soft computing. This hybridization aims to provide more intelligent systems (in terms of performance, parallelism, fault tolerance, adaptivity, and uncertainty management) than the individual one to deal with real-life complex decision making problems.

Both NNs and fuzzy systems work with the principle of adaptivity in the estimation of input-output function without any precise mathematical model. NNs handle numeric and quantitative information while fuzzy systems handle symbolic and qualitative data. Therefore, an integration of neural and fuzzy systems explores the merits of both and enables one to build more intelligent decision making systems. The judicious integration may be viewed another way as: Fuzzy set theoretic models try to mimic human reasoning and the capability of handling uncertainty, whereas NN models attempt to emulate the architecture and information representation scheme of human brain. In other words, if ANN provides the hardware, fuzzy logic provides the software of a system. In the NF paradigm, several pioneering research efforts have been made in the past [25, 49, 58]. NF hybridization is done broadly in two ways:

NNs that are made capable of handling fuzzy information to augment its application domain (named as fuzzy neural networks (FNN)), and fuzzy systems augmented by NNs to enhance some of their characteristics such as flexibility, speed and adaptability (named as neural-fuzzy systems [49, 58]). The details on these methodologies can be found in [58].

In [30], fuzzy neural networks are designed using fuzzy logical operators like AND ($\wedge$), OR ($\vee$), and implication and t-norm. A fuzzy sigmoid function is applied at the neuron. The neuro-fuzzy systems for classification, using multilayer perceptron and self-organizing map, are developed in [42, 57], respectively. While the input vector of the network is defined in terms of linguistic terms, *low*, *medium* and *high* the output vector is expressed as fuzzy membership values. These linguistic terms are basically class independent fuzzy granules. The network is trained based on gradient decent method. These models provide the basic benchmark modules for generating several neuro-fuzzy systems later on. In [15], the network uses fuzzy linguistic variables as its connection weights. The linguistic arithmetic operators like addition, subtraction, multiplication and division are applied to perform operations between any two linguistic variables and thus the weights are modified using linguistic gradient decent method, during training. A fuzzy granular neural network for classifying remote sensing images is provided in [82]. An interval type fuzzy set is chosen in the process of granular neural network for classification in [96]. The network is trained using evolutionary algorithm. A good introduction to neuro-fuzzy inference systems can be found in [31]. A granular neural network for numerical linguistic data fusion, discovering granular knowledge from numerical-linguistic data and converting fuzzy linguistic data into numerical features is defined in [95]. A knowledge based granular neural network for classification is recently described in [20].

1.4.2 Rough Granular Neural Network Models

The principle of granulation has been used in neural networks to design rough neural networks to make them efficient in handling vagueness and incomplete information in the data. In [36], the network contains rough neurons. A rough neuron is a pair of neurons, namely, lower neuron and upper neuron, related to lower and upper values, respectively. The links connecting the lower and upper neurons are initialized with small random real numbers. The logistic function is placed at each of the neurons to control the total output (the sum of product of input and connection weights). Two rule based layer neural networks are developed in [81, 87] where the network parameters are expressed in terms of rough rules. The rules, inducing granules, can be derived from an attribute valued table (information system/decision system). A rough self organizing map for organizing the web users of three educational web sites is developed in [37]. The method uses the notion of interval type of rough set, characterized by the lower and upper approximations of a set. A rough self-organizing map is described in [47]. Here, the network incorporates a set of rough rules as its initial connection weights. The rough rules are obtained from attribute reducts. These are generated, from information system containing three dimensional granular input data, using rough set theory. A method of rough set based radial basis function for classification is provided in [10]. The hidden neurons of radial basis function neural network are set corresponding to the number of features (without requiring cluster information), selected using a novel rough set based method. A rough neural network is designed in [69]. The network consists of two layers, namely input layer and output layer. The outputs of nodes in the input layer (approximation neurons) are rough membership values. These are inputs to the nodes in the output layer (rule based neurons) of the network. The membership values are based on rough membership function [69]. It may be noted that the membership values (initial weights) of the network are not updated. Various other types of rough granular neural networks for solving different types of problems are available in [61].

1.4.3 Rough Fuzzy Granular Neural Network Models

Investigations on integrating rough sets, fuzzy sets and neural networks have been carried out for many years. Different computational methodologies in rough fuzzy-neuro framework are provided in [3, 59]. These are supervised methods. In these frameworks rough set is used in encoding the domain knowledge about the data as network parameters and fuzzy set is utilized in discretization of the feature space. This discretization is accomplished by mapping the conditional attributes of patterns to fuzzy membership values.

In [3], the rough set theoretic parameters determine the size of networks by fixing the number of nodes in hidden layer of networks. Here, the values of attributes are membership values and these are converted as crisp values by applying a threshold to every attribute. Based on the attribute valued table, a rough set methodology is used to generate various attribute reducts. The dependency factors for all the attributes in the reducts are generated and are used as the initial weights of networks.

In [59], a modular sub network, based on multilayer perceptron, is generated corresponding to an attribute reduct, and all such sub modular networks are then integrated to result in a modular evolutionary network. The evolutionary algorithm encodes the intra links of subnetworks with low mutation probability values and the inter links between submodules with high probability values. The judicious integration of rough, fuzzy, neural and evolutionary frameworks helps in achieving superior performance of the network in terms of performance, computation time and network size.

1.5 Pattern Recognition

Pattern recognition from the perspective of human beings is a method of dealing with the activities concerning information processing and taking action subsequently that are normally encountered every day in human life. From the point of machine intelligence, it is a procedure for assessment of a new input to one of a given set of classes. A class is characterized by a set of input patterns having similar properties. A pattern typically consists of a set of features which are essential for recognizing the pattern uniquely. Features represent some measurements characterizing the patterns and are obtained with the help of sensors in various environments. The latter part is called data acquisition. For example, some of the measurements could be in terms of length, width, height, mean, standard deviation, etc. The assignment of the input patterns into one of the classes/categories is done using the tasks like feature selection, classification, clustering, and so on.

A typical pattern recognition system therefore consists of three phases, namely, data acquisition (measurement space), feature selection/extraction (feature space) and classification/clustering (decision space). In the data acquisition phase, depending on the environment within which the objects are to be classified/clustered, data are gathered using a set of sensors. These are then passed on to the feature selection/extraction phase, where the dimensionality of the data is reduced by retaining/measuring only some characteristic features or properties. In a broader perspective, this stage significantly influences the entire recognition process. Finally, in the classification/clustering phase, the selected/extracted features are passed on to the classifying/clustering system that evaluates the incoming information and makes a final decision regarding its class belonging. This phase basically establishes a transformation between the features and the classes/clusters.

1.5.1   Data Acquisition

Pattern recognition techniques are applicable in a wide domain, where the data may be qualitative, quantitative, or both; they may be numerical, linguistic, pictorial, or any combination thereof. The collection of data constitutes the data acquisition phase. Generally, the data structures that are used in pattern recognition systems are of two types: *object data vectors* and *relational data*. Object data, a set of numerical vectors, is represented in the sequel as $Y = \{\mathbf{y}_1, \mathbf{y}_2, \ldots, \mathbf{y}_n\}$, a set of n feature vectors in the p-dimensional measurement space Ω_Y. An sth object, $s = 1, 2, \ldots, n$, observed in the process has vector $\mathbf{y}_s$ as its numerical representation; y_{si} is the ith ($i = 1, 2, \ldots, p$) feature value associated with the sth object. Relational data is a set of n^2 numerical relationships, say $\{r_{sq}\}$, between pairs of objects. In other words, r_{sq} represents the extent to which sth and qth objects are related in the sense of some binary relationship ρ. If the objects that are pairwise related by ρ are called $O = \{o_1, o_2, \ldots, o_n\}$, then $\rho : O \times O \rightarrow I\!R$.

1.5.2 Feature Selection/Extraction

Feature selection/extraction is a process of selecting a map of the form $X = f(Y)$, by which a sample $\mathbf{y}$ $(=[y_1, y_2, \ldots, y_p])$ in a p-dimensional measurement space Ω_Y is transformed into a point $\mathbf{x}$ $(=[x_1, x_2, \ldots, x_{p'}])$ in a p'-dimensional feature space Ω_X, where $p' < p$. The main objective of this task [13] is to retain/generate the optimum salient characteristics necessary for the recognition process and to reduce the dimensionality of the measurement space Ω_Y so that effective and easily computable algorithms can be devised for efficient classification. The problem of feature selection/extraction has two aspects—formulation of a suitable criterion to evaluate the goodness of a feature set and searching the optimal set in terms of the criterion. In general, those features are considered to have optimal saliencies for which interclass/intraclass distances are maximized/minimized. The criterion of a good feature is that it should be unchanging with any other possible variation within a class, while emphasizing differences that are important in discriminating between patterns of different types.

The major mathematical measures so far devised for the estimation of feature quality are mostly statistical in nature, and can be broadly classified into two categories— *feature selection in the measurement space* and *feature selection in a transformed space*. The techniques in the first category generally reduce the dimensionality of the measurement space by discarding redundant or least information carrying features. On the other hand, those in the second category utilize all the information contained in the measurement space to obtain a new transformed space, thereby mapping a higher dimensional pattern to a lower dimensional one. This is referred to as feature extraction.

In last decade, different algorithms for feature selection are developed on the principle of evaluating goodness of every feature and searching the maximum relevance and minimum redundancy among features. The algorithm in [68] is based on mutual information measure where it aims to find a set of features which are more relevant to the mean of its class and mutually different form each other. When two selected features are non-redundant (different from each other and dependency between them is low), then their class discrimination power is high. Some other algorithms include branch and bound, sequential forward and backward search and sequential floating point search methods. An algorithm for feature selection on the basis of feature similarity measure, called maximum information compression index, is described in [41]. The investigations for feature selection have also been carried in the framework of neural networks. A method of multi-layer feed forward network for feature selection is described in [83]. The network is trained by augmented cross entropy error function. A neural network for unsupervised feature selection is defined in [48]. Here, minimization of fuzzy feature evaluation index, based on gradient decent method, is used for training the network.

In microarray data, a feature is represented with a gene. In the framework of soft computing, rough set based method for gene selection is provided in [10]. Here, the method evolves the genes into groups which are evaluated using a validation measure. Out of all the groups of genes, the group which provides the best value for the measure, is selected for further experiments. Methods for dimensionality reduction using fuzzy rough sets are defined in [32, 39, 63]. While the methods in [39] is based on maximization of relevance and minimization of redundancy among selected genes, the algorithms in [32, 63] are based on fuzzy similarity relations. Here, [32, 39] use class information in the formation whereas it is not used in case of [63].

1.5.3 Classification

The problem of classification is basically one of partitioning the feature space into regions, one region for each category of input. Thus it attempts to assign every data point in the entire feature space to one of the possible, say M, classes. In real life, the complete description of the classes is not known. We have instead a finite and usually smaller number of samples which often provides partial information for optimal design of feature selector/extractor or classifying/clustering system. Under such circumstances, it is assumed that these samples are representative of the classes. Such a set of typical patterns is called a *training set*. On the basis of the information gathered from the samples in the training set, the pattern recognition systems are designed; i.e., we determine the values of the parameters for different tasks of pattern recognition. Design of a classification or clustering scheme can be made with labeled or unlabeled data. When the computer is given a set of objects with known classifications (i.e., labels) and is asked to classify an unknown object based on the information acquired by it during training, we call the design scheme *supervised learning*; otherwise we call it *unsupervised learning*. Supervised learning is used for classifying different objects, while clustering is performed through unsupervised learning.

Pattern classification, by its nature, admits many approaches, sometimes complementary, sometimes competing, to provide solution of a given problem. These include *decision theoretic approach* (both *deterministic* and *probabilistic*), *connectionist approach, fuzzy and rough set theoretic approach* and *hybrid or soft computing approach*.

In the decision theoretic approach, once a pattern is transformed, through feature evaluation, to a vector in the feature space, its characteristics are expressed only by a set of numerical values. Classification can be done by using deterministic or probabilistic techniques [17]. In deterministic classification approach, it is assumed that there exists only one unambiguous pattern class corresponding to each of the unknown pattern vectors. Two typical examples in this category are minimum distance classifier [84] and K-NN classifier [12].

In most of the practical problems, the features are usually noisy and the classes in the feature space are overlapping. In order to model such systems, the features $x_1, x_2, \ldots, x_i, \ldots, x_p$ are considered as random variables in the probabilistic approach. The most commonly used classifier in such probabilistic systems is the *Bayes maximum likelihood classifier* [14, 17].

In order to emulate some aspects of the mechanism of human intelligence and discrimination power, the decision-making systems have been made artificially intelligent. Connectionist approaches (or artificial neural network based approaches) to pattern classification are attempts to achieve these goals and have drawn the attention of researchers because of their major characteristics such as adaptivity, robustness/ruggedness, speed and optimality. These classifiers incorporate artificial neurons which are connected to each other in a feed forward fashion. The links connecting the neurons are initialized with random real numbers within 0 to 1. A sigmoid function may be associated with all the neurons, except the input ones, of the networks for computing input output relations. Single layer perceptron, multilayer perceptron and radial basis function networks are examples of the connectionist methods, and are available in [27].

The aforesaid decision theoretic and connectionist approaches to pattern recognition can again be fuzzy set theoretic [51, 93, 94] in order to handle uncertainties, arising from vague, incomplete, linguistic, overlapping patterns, etc., at various stages of pattern recognition systems. Fuzzy set theoretic classification approach is developed based on the realization that a pattern may belong to more than one class, with varying degrees of class membership. Accordingly, fuzzy decision theoretic, fuzzy syntactic, fuzzy neural approaches are developed [6, 35, 44, 51, 56, 92, 94]. On the other hand, the theory of rough sets [64–66] has emerged for managing uncertainty that arises from granularity in the domain of discourse. While the fuzzy set theoretic approach uses the notion of class membership of a pattern, rough set methods revolve around the notion of approximating a set or class or concept from its lower and upper sides over a granular domain. This theory is particularly proven to be useful in the representation of and reasoning with vague and/or imprecise knowledge, data classification, data analysis, machine learning, and knowledge discovery [56, 62, 78].

There have been several attempts over the last few decade to evolve new approaches to pattern recognition and deriving their hybrids by judiciously combining fuzzy logic, artificial neural networks, genetic algorithms and rough set theory, for developing an efficient paradigm called *soft computing* [91]. Here integration is done in a cooperative, rather than a competitive, manner. The result is a more intelligent and robust system [3, 40, 56, 62] providing a human-interpretable, low cost, approximate solution, as compared to traditional techniques.

Other types of classifiers like decision tree based algorithms [77], support vector machine (SVM) [79], and genetic algorithms [2, 7, 26] are also available.

1.5.4 Clustering

Clustering refers to partitioning of patterns in the feature space into meaningful homogeneous groups. The word meaningful is subjective. Clustering is an unsupervised classification method which does not require the class information of the patterns for determining the clusters. The clustering methods can be categorized as follows:

(1) *Partitioning methods*: Given a data set of n objects and k, the number of clusters to form, a partitioning algorithm organizes the objects into k partitions, where each partition represents a cluster. The clusters are formed to optimize an objective partitioning criterion, often called a *similarity function*, such as distance, so that the objects within a cluster are 'similar,' whereas the objects of different clusters are 'dissimilar.' A partitioning method starts with an initial partition and uses an iterative refinement technique that attempts to improve the partitioning by moving objects from one group to another. The most well-known and commonly used partitioning methods are k-means, k-medoids and their variations. Probabilistic clustering methods, often based on the mixture modeling and expectation maximization (EM) algorithm, are also popular. Partitional methods work well for finding spherical shaped clusters in small to medium-sized data sets. For clustering very large data sets and to find clusters with complex shapes, these methods need to be extended.

(2) *Hierarchical clustering methods*: The hierarchical clustering methods [33] constitute a dendrogram (a hierarchical binary tree) using any one of distance metrics among single linkage, complete linkage and average linkage in graph theoretic framework. The distance between two clusters is computed by either the value of shortest link between two patterns in two different clusters (single linkage), or the value of longest link between two patterns in two different clusters (complete linkage) or the average of distances between all objects in the first cluster and all objects in the second cluster (average linkage). Hierarchical clustering methods are of two types: (i) Agglomerative (ii) Divisive. The first one is a bottom up approach. This approach starts with every pattern as a single cluster and merges a pair of clusters into one cluster, according to the closest distance between the pair. The process of merging the clusters is terminated when all the clusters are combined into a single cluster. The second method is top down approach. This method starts with all patterns in one cluster and split the single cluster into two clusters such that one cluster is dissimilar to other. The method continues the process of splitting the clusters till each pattern forms a single cluster. Hierarchical methods suffer from the fact that once a step (merge or split) is done, it can never be undone. This rigidity leads to sensitivity to noise in the data.

Both the partitioning and graph theoretic approaches have their advantages and disadvantages and cannot directly be applied for data mining. While the partitioning schemes are fast and easily scalable to large databases, they can produce only convex clusters and are sensitive to initialization of the parameters. The hierarchical methods can model arbitrary shaped clusters but are slow and sensitive to noise. It may be noted that the advantages of one are complementary in overcoming the limitations of the other.

(3) *Density based methods*: Clustering algorithm have been also developed based on the notion of density. The general idea is to continue growing the given cluster as long as the density (number of data points) in the 'neighborhood' exceeds some threshold. Such a method can be used to filter out noise and discover clusters of

arbitrary shape. The method requires the parameters like radius, minimum number of patterns and a core pattern. A pattern is called a core if it has greater than or equal to the minimum number of patterns, as specified before, within its neighborhood. A pattern is said to be within radius of a core pattern if the distance from the core pattern to its neighborhood pattern is less than or equal to the epsilon radius, defined by the user. The methods work on the principle of density reachable where a pattern s is density reachable from a pattern r, when r is a core and s is within the epsilon neighborhood of r. DBSCAN [18], OPTICS [1], and DENCLUE [29] are typical examples in this class.

(4) *Grid-based methods*: Grid-based methods quantize the object space into a finite number of cells (granules) that form a grid structure. The clustering operations are performed on the grid structure (i.e., on the granulated space). The main advantage of this approach is its fast processing time, which is typically independent of the number of data objects.

Besides the above mentioned methods other clustering algorithms based on neural networks and granular neural networks are also developed. Self-organizing map (SOM) [34] and adaptive neural networks (ART) [8] are examples of this category. Clustering methodologies in granular neural networks framework are provided in [28, 37, 47] where the granulation is developed using fuzzy and rough sets. Information granulation based on rough reducts have also been used [54] in determining the initial number clusters, k, in EM algorithm of partitional methods of clustering.

It may be mentioned here that the challenges in classification and clustering are different. In the former case, only a fraction of the labeled patterns (patterns with class information), called training set, is provided and one is asked to estimate the class label of the remaining patterns. In the latter case, on the other hand, the entire data set with no class information is given, and the task is to partition them into meaningful homogeneous pockets.

1.6 Data Mining and Soft Computing

Data mining refers to the process of discovering or mining of knowledge/patterns from databases. Here, databases may include relational databases, data warehouses, data streams, world wide web data, text data, microarray data, real life data, and multimedia data. Different tools, methods and applications of neural networks, fuzzy sets, rough sets, granular neural networks, etc., can be applied on such databases in finding relations among patterns at various levels of abstraction for discovery of knowledge.

From pattern recognition perspective, data mining can be viewed as applying pattern recognition and machine learning principles to very large (in terms of both size and dimension) and heterogeneous data sets. And

$$Data\ Mining + Knowledge\ interpretation \equiv Knowledge\ Discovery. \qquad (1.20)$$

That means, knowledge discovery in database (KDD) is a process of identifying valid, novel, potentially useful, and ultimately understandable patterns in data. Variety of algorithms for mining data using clustering and feature selection in pattern recognition and soft computing perspective are available in [56].

As seen, the aim of data mining methodologies is to explore relations among the patterns in a representational form and meaningful manner. The methodologies are aimed to be flexible for processing information in ambiguous situations arising in real life databases. In this regard new methodologies can be developed by integrating fuzzy sets, fuzzy rough sets and neural networks in soft computing paradigm for handling real life data sets.

Among the real life data sets, the most notable one used in this book is the microarray gene expression data set which is helpful in predicting genes related with either disease (diseased microarray data) or any particular biological function (functional microarray data). The DNA microarray is made by a thin glass or nylon substrates or oligonucleotide arrays that contain thousands of spots representing genes from different samples. mRNAs are extracted from tumor or normal cells (samples) and are converted into tumor cDNAs or normal cDNAs using reverse transcription. A scanner is used to measure the fluorescence values of tumor cDNA, normal cDNA and complementary sequences of DNA and those values are used as label values for different genes. The experimental sections in this book show that mining of patterns from publicly available certain microarray gene expression data sets can reveal the knowledge available with the genes.

1.7 Big Data Issues

In recent years, almost every research field such as bioinformatics, genomics, image processing and online social networks are producing data in huge volume. These data show all the characteristics of Big data (e.g., high volume, high velocity and hight variety) where scalability is one of the important issues for analysis and fruitful conclusion. Chapter 1 of the recently published volume by Pal and Pal [44] addresses the Big data issues from the point of pattern recognition and data mining.

In case of Big data analysis, fuzzy rough granular networks (FRGN) can have the following two advantages over the conventional fuzzy rough models. First, in FRGN, the properties of a data point are embedded with the granule constructed around it. If an algorithm demands to work on part of a data set rather than the complete one then one may find it convenient to use the algorithm on only those granules which represent the partial data and yet can analyze the data from the granular characteristics. Even for the global property analysis, for reducing the execution time of data processing one may restrict the number of granules based on users choice. Experimentally, researchers [50] have found that with the reduction in number of granules within a data set, the rate of improvement in execution time for an algorithm is exponential while the rate of drop in accuracy is linear. Second, granules may be more effectively used in asynchronous distributed systems where one computing unit will only deal with a subset of granules. One may note that, distributed computing system is one of those frameworks which provides a platform for Big data handling. In a distributed system, communication between elements is inherently asynchronous. There is no global clock nor consistent clock rate. Each computer processes independently of others. Some computers in the system have fast clock cycles while others have slower clock cycles. Even if the time was precisely the same on every element in the distributed system, each element would still process the communication at different rates, thus making the commutation asynchronous and providing an opportunity to consider each element as a granule.

1.8 Scope of the Book

The book addresses different models of granular neural networks for pattern recognition and mining tasks along with extensive experimental results on different kinds of real life data sets. The design aspects of the networks, involving the theories of rough set and fuzzy set, and their characteristics are described. The results of the networks for classification, clustering and feature selection/gene selection are compared with those of related and well known classical algorithms. Problems considered include spam email identification, speech recognition, glass type classification, cancer identification, and microarray gene expression analysis. Each of the chapters in this book is organized as follows:

Chapter 2 deals with problem of classification of patterns using fuzzy rough granular neural networks. Some of the existing classical and related methods are described. These are followed by some recent methods developed in [19, 21], based on the multilayer perceptron using a back-propagation algorithm. We provide the development strategy of the network mainly based upon the input vector, initial connection weights determined by fuzzy rough set theoretic concepts, and the target vector. While the input vector is described in terms of fuzzy granules, the target vector is expressed in terms of fuzzy class membership values and zeros. Crude domain knowledge about the input data is represented in the form of a decision table (attribute valued table) which is divided into sub tables corresponding to the number of classes. The data in each decision table is converted into granular form. The syntax of these decision tables automatically determines the appropriate number of hidden nodes, while the dependency factors from all the decision tables are used as initial weights. The dependency factor of each attribute and the average degree of the dependency factor of all the attributes with respect to decision classes are considered as initial connection weights between the nodes of the input layer and the hidden layer, and the hidden layer and the output layer, respectively. The effectiveness of the FRGNN1 and FRGNN2 in classification of fuzzy patterns is demonstrated on real-life data sets.

Chapter 3 describes a fuzzy rough granular self-organizing map (FRGSOM) [22, 23] along with some classical clustering techniques and related methodologies based on self-organizing map. Emphasis is given on FRGSOM as it is the first attempt in integrating fuzzy, rough and granular concepts with SOM. The FRGSOM involves a 3-dimensional (3D) linguistic vector and connection weights, defined in an unsupervised manner, for clustering patterns having overlapping regions. Each feature of a pattern is transformed into 3D granular space using a π-membership function with centers and scaling factors, corresponding to the linguistic terms *low*, *medium* or *high*. The three dimensional linguistic vectors are then used to develop granulation structures, based on a user defined α-value. The granulation structures are presented to decision table as its crisp decision classes. The data in decision table is used to determine the dependency factors of the conditional attributes using the concept of fuzzy rough sets. The dependency factors are used as initial connection weights of the FRGSOM. The FRGSOM is then trained through a competitive learning of the self-organizing map. A new fuzzy rough entropy measure is defined [23], based on the resulting clusters and using the concept of fuzzy rough sets. The effectiveness of the FRGSOM and the utility of "fuzzy rough entropy" in evaluating cluster quality are demonstrated on different real life data sets, including microarrays, with varying dimensions.

Chapter 4 deals with the task for identifying salient features of data. In this regard, some existing methodologies for feature selection are first briefly described and then a recent technique using a granular neural network (FRGNN3) [24], based on the concepts of fuzzy set and a newly defined fuzzy rough set, is explained in details. The formation of FRGNN3 mainly involves an input vector, initial connection weights and a target value. Each feature of the data is normalized between 0 and 1 and used to develop granulation structures by a user defined α-value. The input vector and the target value of the network are defined using granulation structures, based on the concept of fuzzy sets. The same granulation structures are also presented to a decision system. The decision system helps in extracting the domain knowledge about data in the form of dependency factors, using the notion of new fuzzy rough set. These dependency factors are assigned as the initial connection weights of the network. It is then trained using minimization of a novel feature evaluation index in an unsupervised manner. The effectiveness of the granular neural network, in evaluating selected features, is demonstrated on different data sets.

In Chap. 5 different clustering techniques based on self-organizing map and other partitional methods are first mentioned. These are followed by a recent technique, called granular self-organizing map (GSOM) [76], which is developed by integrating the concept of a fuzzy rough set with the self-organizing map (SOM). While training the GSOM, the weights of a winning neuron and the neurons lying within its neighborhood are updated through a modified learning procedure of SOM. The neighborhood is newly developed using fuzzy rough sets. The clusters (granules), evolved by GSOM, are presented to a decision table as its decision classes. Based on the decision table, a method of gene selection is developed. The effectiveness of the GSOM is shown in both clustering samples and developing an unsupervised fuzzy rough feature selection (UFRFS) method for gene selection in microarray data. While the superior results of the GSOM, as compared to related clustering methods, are provided in terms of β-index, DB-index and fuzzy rough entropy, the genes selected by UFRFS are not only better in terms of classification accuracy and a feature evaluation index than the related methods.

Chapter 6 addresses the problem of gene analysis using their expression values. Here, the different tasks of gene expression analysis like preprocessing, distance measures, gene clustering and ordering and gene function prediction are described. Emphasis is given on the issues where functions of unclassified genes are predicted from gene expressions by applying optimal gene ordering mechanism using genetic algorithm based FRAG_GALK [74], and a hybrid approach where FRAG_GALK is used for gene ordering in each cluster obtained from solutions of partitional clustering methods [73]. Finally, a method for gene function prediction through integration of multiple biological data sources is demonstrated [75]. The data sources are gene expression, phenotypic profiles, protein sequences, Kyoto Encyclopedia of Genes and Genomes (KEGG) pathway information and protein-protein interaction data. Experimental results indicated that the combination of FRAG_GALK and partitional clustering is a promising tool for microarray gene expression analysis involving gene function prediction. Further, the technique for integrating multiple biological data sources made it possible to predict the function of 12 unclassified yeast genes for the first time. It may be noted that, granular computing is not utilized in the methods described in this chapter, and it mainly deals with combinatorial optimization and genetic algorithm, a component of soft computing.

Chapter 7 deals with the prediction of RNA structure which is invaluable in creating new drugs and understanding genetic diseases. Several deterministic algorithms and soft computing based techniques, such as genetic algorithms, artificial neural networks, and fuzzy logic, have been developed for more than a decade to determine the structure from a known RNA sequence. It is important to mention that although the aforementioned basic soft computing techniques are used for RNA secondary structure prediction, rough sets and granular computing are still not explored for this purpose. Soft computing gained importance with the need to get approximate solutions for RNA sequences by considering the issues related with kinetic effects, cotranscriptional folding and estimation of certain energy parameters. A brief description of some of the soft computing based techniques developed for RNA secondary structure prediction, is presented along with their relevance. The basic concepts of RNA and its different structural elements like helix, bulge, hairpin loop, internal loop, and multiloop are described. These are followed by different methodologies, employing genetic algorithms, artificial neural networks and fuzzy logic. The role of various metaheuristics, like simulated annealing, particle swarm optimization, ant colony optimization and tabu search is also discussed. A relative comparison among different techniques, in predicting 12 known RNA secondary structures, is presented, as an example. Future challenging issues are then mentioned.

Appendix provides a brief description of various data sets used in the aforesaid experimental investigations.

References

1. Ankerst, M., Breunig, M.M., Kriegel, H.P., Sander, J.: OPTICS: ordering points to identify the clustering structure. In: Proceedings of ACM International Conference Management of Data, pp. 49–60 (1999)
2. Bandyopadhyay, S., Pal, S.K.: Classification and Learning Using Genetic Algorithms: Applications in Bioinformatics and Web Intelligence (Natural Computing Series). Springer, Heidelberg (2007)
3. Banerjee, M., Mitra, S., Pal, S.K.: Rough fuzzy MLP: knowledge encoding and classification. IEEE Trans. Neural Netw. 9(6), 1203–1216 (1998)
4. Banerjee, M., Pal, S.K.: Roughness of a fuzzy set. Inf. Sci. 93(3–4), 235–246 (1996)
5. Bellman, R.E., Kalaba, R., Zadeh, L.A.: Abstraction and pattern classification. J. Math. Anal. Appl. 13, 1–7 (1966)
6. Bezdek, J.C., Pal, S.K.: Fuzzy Models for Pattern Recognition: Methods that Search for Structures in Data. IEEE Press, New York (1992)
7. Booker, L.B., Goldberg, D.E., Holland, J.H.: Classifier systems and genetic algorithms. Artif. Intell. 40, 235–282 (1989)
8. Carpenter, G.A., Grossberg, S., Rosen, D.B.: ART 2-A: an adaptive resonance algorithm for rapid category learning and recognition. Neural Netw. 4, 493–504 (1991)
9. Cerro, L.F.D., Prade, H.: Rough sets, two fold fuzzy sets and logic. In: Di Nola, A., Ventre, A.G.S. (eds.) Fuzziness in Indiscernibility and Partial Information, pp. 103–120. Springer, Berlin (1986)
10. Chiang, J.H., Ho, S.H.: A combination of rough-based feature selection and RBF neural network for classification using gene expression data. IEEE Trans. Nanobiosci. 7(1), 91–99 (2008)
11. Cornelis, C., Jensen, R., Hurtado, G., Slezak, D.: Attribute selection with fuzzy decision reducts. Inf. Sci. 180(2), 209–224 (2010)
12. Cover, T.M., Hart, P.E.: Nearest neighbor pattern classification. IEEE Trans. Inf. Theor. 13(1), 21–27 (1967)
13. Devijver, P.A., Kittler, J.: Pattern Recognition: A Statistical Approach. Prentice Hall, Englewood Cliffs (1982)
14. Devroye, L., Gyorfi, L., Lugosi, G.: A Probabilistic Theory of Pattern Recognition. Springer, Heidelberg (1996)
15. Dick, S., Kandel, A.: Granular computing in neural networks. In: Pedrycz, W. (ed.) Granular Computing: An Emerging Paradigm, pp. 275–305. Physica Verlag, Heidelberg (2001)
16. Dubois, D., Prade, H.: Rough fuzzy sets and fuzzy rough sets. Int. J. Gen. Syst. 17(2–3), 91–209 (1990)
17. Duda, R.O., Hart, P.E., Stork, D.G.: Pattern Classification, 2nd edn. Wiley-Interscience, New York (2000)
18. Ester, M., Kriegel, H.P., Sander, J., Xu, X.: A density-based algorithm for discovering clusters in large spatial databases with noise. In: Proceedings of 2nd International Conference on Knowledge Discovery and Data Mining (KDD 1996), pp. 226–231 (1996)
19. Ganivada, A., Dutta, S., Pal, S.K.: Fuzzy rough granular neural networks, fuzzy granules, and classification. Theoret. Comput. Sci. 412, 5834–5853 (2011)
20. Ganivada, A., Pal, S.K.: Robust granular neural networks, fuzzy granules and classification. In: Proceedings of 5th International Conference on Rough Sets and Knowledge Technology, pp. 220–227 (2010)
21. Ganivada, A., Pal, S.K.: A novel fuzzy rough granular neural network for classification. Int. J. Comput. Intell. Syst. 4(5), 1042–1051 (2011)
22. Ganivada, A., Ray, S.S., Pal, S.K.: Fuzzy rough granular self organizing map. In: Proceedings of 6th International Conference on Rough Sets and Knowledge Technology, pp. 659–668 (2011)
23. Ganivada, A., Ray, S.S., Pal, S.K.: Fuzzy rough granular self-organizing map and fuzzy rough entropy. Theoret. Comput. Sci. 466, 37–63 (2012)
24. Ganivada, A., Ray, S.S., Pal, S.K.: Fuzzy rough sets, and a granular neural network for unsupervised feature selection. Neural Netw. 48, 91–108 (2013)

25. Ghosh, A., Shankar, B.U., Meher, S.: A novel approach to neuro-fuzzy classification. Pattern Recogn. **22**, 100–109 (2009)
26. Goldberg, D.: Genetic Algorithms in Optimization, Search, and Machine Learning. Addison Wesley, Reading (1989)
27. Haykin, S.: Neural Networks: A Comprehensive Foundation, 2nd edn. Prentice Hall, Upper Saddle River (1998)
28. Herbert, J.P., Yao, J.T.: A granular computing frame work for self-organizing maps. Neuro-computing **72**(13–15), 2865–2872 (2009)
29. Hinneburg, A., Keim, D.A.: An efficient approach to clustering in large multimedia databases with noise. In: Proceedings of 4th International Conference on Knowledge Discovery and Data Mining, pp. 58–65 (1998)
30. Hirota, K., Pedrycz, W.: OR/AND neuron in modeling fuzzy set connectives. IEEE Trans. Fuzzy Syst. **2**(2), 151–161 (1994)
31. Jang, J.S.R.: ANFIS: adaptive-network-based fuzzy inference systems. IEEE Trans. Syst. Man Cybern. **23**(3), 665–685 (1993)
32. Jensen, R., Shen, Q.: New approaches to fuzzy-rough feature selection. IEEE Trans. Fuzzy Syst. **17**(4), 824–838 (2009)
33. Johnson, S.C.: Hierarchical clustering schemes. Psychometrika **32**(3), 241–254 (1967)
34. Kohonen, T.: Self-organizing maps. Proc. IEEE **78**(9), 1464–1480 (1990)
35. Kuncheva, L.I.: Fuzzy Classifier Design. Springer, Heidelberg (2000)
36. Lingras, P.: Rough neural networks. In: Proceedings of 6th International Conference on Information Processing and Management of Uncertainty, pp. 1445–1450 (1996)
37. Lingras, P., Hogo, M., Snorek, M.: Interval set clustering of web users using modified Kohonen self-organizing maps based on the properties of rough sets. Web Intell. Agent Syst. **2**(3), 217–225 (2004)
38. Lingras, P., Jensen, R.: Survey of rough and fuzzy hybridization. In: Proceedings of IEEE International Conference on Fuzzy Systems, pp. 1–6 (2007)
39. Maji, P., Pal, S.K.: Fuzzy-rough sets for information measures and selection of relevant genes from microarray data. IEEE Trans. Syst. Man. Cybern. Part B **40**(3), 741–752 (2010)
40. Maji, P., Pal, S.K.: Rough-Fuzzy Pattern Recognition: Applications in Bioinformatics and Medical Imaging. Wiley, Hoboken (2011)
41. Mitra, P., Murthy, C.A., Pal, S.K.: Unsupervised feature selection using feature similarity. IEEE Trans. Pattern Anal. Mach. Intell. **24**(4), 301–312 (2002)
42. Mitra, S., Pal, S.K.: Self-organizing neural network as a fuzzy classifier. IEEE Trans. Syst. Man Cybern. **24**(3), 385–399 (1994)
43. Ouyang, Y., Wang, Z., Zhang, H.P.: On fuzzy rough sets based on tolerance relations. Inf. Sci. **180**(4), 532–542 (2010)
44. Pal, A., Pal, S.K.: Pattern Recognition and Big Data (2017)
45. Pal, S.K.: Computational theory perception (CTP), rough-fuzzy uncertainty analysis and mining in bioinformatics and web intelligence: a unified framework. In: Transactions on Rough Set. LNCS, vol. 5946, pp. 106–129 (2009)
46. Pal, S.K.: Granular mining and rough-fuzzy pattern recognition: a way to natural computation. IEEE Intell. Inf. Bull. **13**(1), 3–13 (2012)
47. Pal, S.K., Dasgupta, B., Mitra, P.: Rough self-organizing map. Appl. Intell. **21**(1), 289–299 (2004)
48. Pal, S.K., De, R.K., Basak, J.: Unsupervised feature evaluation: a neuro-fuzzy approach. IEEE Trans. Neural Netw. **11**(2), 366–376 (2000)
49. Pal, S.K., Ghosh, A.: Neuro-fuzzy computing for image processing and pattern recognition. Int. J. Syst. Sci. **27**, 1179–1193 (1996)
50. Pal, S.K., Kundu, S.: Granular social network: model and applications. In: Zomaya, A., Sakr, S. (eds.) Handbook of Big Data Technologies. Springer, Heidelberg (2017, to appear)
51. Pal, S.K., Majumder, D.D.: Fuzzy Mathematical Approach to Pattern Recognition. Wiley, New York (1986)

52. Pal, S.K., Meher, S.K.: Natural computing: a problem solving paradigm with granular information processing. Appl. Soft Comput. **13**(9), 3944–3955 (2013)
53. Pal, S.K., Meher, S.K., Dutta, S.: Class-dependent rough-fuzzy granular space, dispersion index and classification. Pattern Recogn. **45**(7), 2690–2707 (2012)
54. Pal, S.K., Mitra, P.: Multispectral image segmentation using rough set initialized EM algorithm. IEEE Trans. Geosci. Remote Sens. **40**(11), 2495–2501 (2002)
55. Pal, S.K., Mitra, P.: Case generation using rough sets with fuzzy representation. IEEE Trans. Knowl. Data Eng. **16**(3), 292–300 (2004)
56. Pal, S.K., Mitra, P.: Pattern Recognition Algorithms for Data Mining. Chapman and Hall/CRC, Boca Raton (2004)
57. Pal, S.K., Mitra, S.: Multilayer perceptron, fuzzy sets and classification. IEEE Trans. Neural Netw. **3**(5), 683–697 (1992)
58. Pal, S.K., Mitra, S.: Neuro-Fuzzy Pattern Recognition: Methods in Soft Computing. Wiley-Interscience, New York (1999)
59. Pal, S.K., Mitra, S., Mitra, P.: Rough fuzzy MLP: modular evolution, rule generation and evaluation. IEEE Trans. Knowl. Data Eng. **15**(1), 14–25 (2003)
60. Pal, S.K., Peters, J.F.: Rough Fuzzy Image Analysis: Foundations and Methodologies. Chapman and Hall/CRC, Boca Raton (2010)
61. Pal, S.K., Polkowski, L., Skowron, A.: Rough-Neural Computing: Techniques for Computing with Words. Springer, Heidelberg (2004)
62. Pal, S.K., Skowron, A.: Rough-Fuzzy Hybridization: A New Trend in Decision Making. Springer, New York (1999)
63. Parthalain, N.M., Jensen, R.: Unsupervised fuzzy-rough set-based dimensionality reduction. Inf. Sci. **229**(3), 106–121 (2013)
64. Pawlak, Z.: Rough Sets: Theoretical Aspects of Reasoning About Data. Kluwer Academic Publishers, Dordrecht (1992)
65. Pawlak, Z., Skowron, A.: Rough sets and Boolean reasoning. Inf. Sci. **177**(3), 41–73 (2007)
66. Pawlak, Z., Skowron, A.: Rough sets: some extensions. Inf. Sci. **177**(3), 28–40 (2007)
67. Pedrycz, W., Vukovich, G.: Granular neural networks. Neurocomputing **36**, 205–224 (2001)
68. Peng, H.C., Long, F., Ding, C.: Feature selection based on mutual information: criteria of max-dependency, max-relevance, and min-redundancy. IEEE Trans. Pattern Anal. Mach. Intell. **27**(8), 1226–1238 (2005)
69. Peters, J.F., Skowron, A., Han, L., Ramanna, S.: Towards rough neural computing based on rough membership functions: theory and application. In: Ziarko, W., Yao, Y. (eds.) Rough Sets and Current Trends in Computing, pp. 611–618. Springer, Heidelberg (2001)
70. Polkowski, L., Skowron, A.: Rough mereology: a new paradigm for approximate reasoning. Int. J. Approx. Reason. **15**, 333–365 (1996)
71. Polkowski, L., Tsumoto, S., Lin, T.Y.: Rough Set Methods and Applications. Physica, Heidelberg (2001)
72. Qian, Y., Liang, J., Yao, Y., Dang, C.: MGRS: a multi-granulation rough set. Inf. Sci. **180**, 949–970 (2010)
73. Ray, S.S., Bandyopadhyay, S., Pal, S.K.: Gene ordering in partitive clustering using microarray expressions. J. Biosci. **32**(5), 1019–1025 (2007)
74. Ray, S.S., Bandyopadhyay, S., Pal, S.K.: Genetic operators for combinatorial optimization in TSP and microarray gene ordering. Appl. Intell. **26**(3), 183–195 (2007)
75. Ray, S.S., Bandyopadhyay, S., Pal, S.K.: Combining multi-source information through functional annotation based weighting: gene function prediction in yeast. IEEE Trans. Biomed. Eng. **56**(2), 229–236 (2009)
76. Ray, S.S., Ganivada, A., Pal, S.K.: A granular self-organizing map for clustering and gene selection in microarray data. IEEE Trans. Neural Netw. Learn. Syst. **27**(9), 1890–1906 (2016)
77. Safavian, S.R., Landgrebe, D.: A survey of decision tree classifier methodology. IEEE Trans. Syst. Man Cybern. **21**(3), 660–674 (1991)
78. Saha, S., Murthy, C.A., Pal, S.K.: Rough set based ensemble classifier for web page classification. Fundamenta Informaticae **76**(1–2), 171–187 (2007)

79. Schölkopf, B., Smola, A.J.: Learning with Kernels: Support Vector Machines, Regularization, Optimization, and Beyond. MIT Press, Cambridge (2001)
80. Sen, D., Pal, S.K.: Generalized rough sets, entropy and image ambiguity measures. IEEE Trans. Syst. Man Cybern. Part B **39**(1), 117–128 (2009)
81. Szczuka, M.: Refining classifiers with neural networks. Int. J. Comput. Inf. Sci. **16**(1), 39–55 (2001)
82. Vasilakos, A., Stathakis, D.: Granular neural networks for land use classification. Soft Comput. **9**(5), 332–340 (2005)
83. Verikas, A., Bacauskiene, M.: Feature selection with neural networks. Pattern Recogn. Lett. **23**(11), 1323–1335 (2002)
84. Wacker, A.G.: Minimum distance approach to classification. Ph.D. thesis, Purdue University, Lafayette, Indiana (1972)
85. Yang, X., Song, X., Dou, H., Yang, J.: Multigranulation rough set: from crisp to fuzzy case. Ann. Fuzzy Math. Inf. **1**, 55–70 (2011)
86. Yao, Y.Y.: A partition model of granular computing. In: Transactions on Rough Sets. LNCS, vol. 3100, pp. 232–253 (2004)
87. Yasdi, R.: Combining rough sets and neural learning method to deal with uncertain and imprecise information. Neurocomputing **7**(1), 61–84 (1995)
88. Zadeh, L.A.: Fuzzy sets. Inf. Control **8**, 338–353 (1965)
89. Zadeh, L.A.: Outline of a new approach to the analysis of complex systems and decision processes. IEEE Trans. Syst. Man Cybern. SMC **3**, 28–44 (1973)
90. Zadeh, L.A.: The concept of a linguistic variable and its application to approximate reasoning-II. Inf. Sci. **8**, 301–357 (1975)
91. Zadeh, L.A.: Fuzzy logic, neural networks, and soft computing. Commun. ACM **37**(3), 77–84 (1994)
92. Zadeh, L.A.: Fuzzy logic = computing with words. IEEE Trans. Fuzzy Syst. **4**, 103–111 (1996)
93. Zadeh, L.A.: Towards a theory of fuzzy information granulation and its centrality in human reasoning and fuzzy logic. Fuzzy Sets Syst. **90**, 111–127 (1997)
94. Zadeh, L.A.: A new direction in AI: towards a computational theory of perceptions. AI Mag. **22**, 73–84 (2001)
95. Zhang, Y.Q., Fraser, M.D., Gagliano, R., Kandel, A.: Granular neural networks for numerical-linguistic data fusion and knowledge discovery. IEEE Trans. Neural Netw. **11**(3), 658–667 (2000)
96. Zhang, Y.Q., Jin, B., Tang, Y.: Granular neural networks with evolutionary interval learning. IEEE Trans. Fuzzy Syst. **16**(2), 309–319 (2008)
97. Zhu, W., Wang, F.Y.: On three types of covering-based rough sets. IEEE Trans. Knowl. Data Eng. **19**(8), 1649–1667 (2007)

Chapter 2
Classification Using Fuzzy Rough Granular Neural Networks

2.1 Introduction

In Chap. 1, a comprehensive description of granular computing, rough fuzzy computing and different classification, clustering and feature selection methods is provided. This chapter deals with pattern classification in the framework of granular neural networks. As mentioned in Chap. 1, the problem of classification is basically one of partitioning the feature space into regions, one region for each category of input. A finite and usually a small number of samples, called a *training set*, provides partial information for optimal design of classifying system, in terms of the values of the parameters, for different tasks of pattern recognition. Different types of classifiers are briefly described in Sect. 1.5.3 of Chap. 1.

As mentioned there, integration of various components of soft computing like fuzzy sets, rough sets, fuzzy rough sets and artificial neural networks provides a new way in which the fuzzy neural networks and rough fuzzy neural networks for classification are developed in pattern recognition framework. These methods can handle uncertainties arising from overlapping classes. In doing so, fuzzy sets are associated with multilayer perceptron (MLP) to design neuro-fuzzy inference model [6]. A fuzzy MLP for classification is designed in [15]. Here fuzzy sets are used to define input vectors of MLP in terms of 3-dimensional (3D) linguistic terms corresponding to *low*, *medium* and *high*.

In this chapter we first describe the salient features of neuro-fuzzy inference model and fuzzy MLP. The networks subsequently developed based on these models are then discussed. These are followed by a rough fuzzy neural network, by integrating rough set theory with fuzzy MLP, along with its characteristic features. Incorporation of rough reducts, as obtained from the training data, as initial network parameters is seen to enhance the classification performance while reducing training time significantly. Finally we describe in detail some of the recent granular neural networks, namely, FRGNN1 [4] and FRGNN2 [5] with experimental results.

© Springer International Publishing AG 2017   39
S.K. Pal et al., *Granular Neural Networks, Pattern Recognition
and Bioinformatics*, Studies in Computational Intelligence 712,
DOI 10.1007/978-3-319-57115-7_2

2.2 Adaptive-Network-Based Fuzzy Inference System

The adaptive-network-based fuzzy inference system (ANFIS), as described in [6], has two inputs x and y and one output z. It is assumed that the network consists of two fuzzy if-then rules of Takagi and Sugenos type, namely,

Rule 1: If x is A_1 and y is B_1, then $f_1 = p_1x + q_1y + r_1$,

Rule 2: If x is A_2 and y is B_2, then $f_2 = p_2x + q_2y + r_2$.

The architecture of ANFIS is available in [6]. The node functions in the layers are described as follows:

Layer 1: Every node i in this layer is a square node with a node function

$$O_i^1 = \mu_{A_i}(x) \tag{2.1}$$

where x is the input to node i, and A is the linguistic label like, small and large, associated with this node function. O_i^1 is the output of the membership function of A_i and it specifies the degree to which the given x satisfies the quantifier A_i. Here bell shaped membership function with maximum 1 and minimum 0 is used.

Layer 2: Every node in this layer is a circle node labeled with $\prod$ which multiplies the incoming signals (input values) and sends the product as output signal. For instance,

$$w_i = \mu_{A_i}(x) \times \mu_{B_i}(y), i = 1 \text{ and } 2. \tag{2.2}$$

Each node output represents the firing strength of a rule. A T-norm operator, that performs generalized AND, can be used as the node function in this layer.

Layer 3: Every node in this layer is a circle node labeled with N. The ith node calculates the ratio of the ith rules firing strength to the sum of all rules firing strengths.

$$\overline{w_i} = \frac{w_i}{w_1 + w_2}, i = 1 text and 2. \tag{2.3}$$

Outputs of this layer are called normalized firing strengths.

Layer 4: Every node i in this layer is a square node with a node function

$$O_i^4 = \overline{w_i} f_i = \overline{w_i}(p_i x + q_i y + r_i), \tag{2.4}$$

where $\overline{w_i}$ is the output of layer 3, and $\{p_i, q_i, r_i\}$ is a parameter set.

Layer 5: The single node in this layer is a circle node labeled with $\sum$, that computes the overall output, summation of all incoming signals,

$$O_1^5 = overall output = \sum_i \overline{w_i} f_i = \frac{\sum_i w_i f_i}{\sum_i w_i} \tag{2.5}$$

A learning algorithm of ANFIS involves forward pass and backward pass. In the forward pass, functional signals go forward till layer 4 and the parameters are identified by least square estimate as in [6]. In the backward pass, the error rates propagate backward and the parameters are updated using the gradient descent method. The details of the learning algorithm are elaborated in [6].

2.3 Fuzzy Multi-layer Perceptron

This is a pioneering model where the integration of fuzzy sets and multilayer perceptron (MLP) resulted in the fuzzy MLP (FMLP), described in [15], for classification. The model involves fuzzy input vector and fuzzy output vector. The fuzzy input vector consists of 3-dimensional feature values whereas fuzzy output vector has membership values corresponding to different classes. Each feature is represented as a 3-dimensional membership vector using Π-membership functions, corresponding to the linguistic terms *low*, *medium* and *high*. The values of parameters for the Π-membership function, corresponding to the three linguistic terms, are chosen as in [15]. Network parameters are updated using the back propagation algorithm depending on the membership values at output nodes. The model provides a judicious integration of the capabilities of MLP in generating non linear boundaries and of fuzzy set in handling uncertainties arising from overlapping regions. Unlike, MLP, FMLP can accept both numerical and linguistic input. The superiority of FMLP over MLP in terms of classification performance and learning time has been adequately established. In Sect. 2.6, FMLP is used to design the granular models FRGNN1 [4] and FRGNN2 [5] with random weights within 0–1.

2.4 Knowledge Based Fuzzy Multi-layer Perceptron

A knowledge based fuzzy MLP for classification and rule generation is developed in [10]. Here, the number of hidden nodes and the connection weights of the network are determined by considering the values of the 3-dimensional features within an interval and outside the interval (compliment of the interval) for a particular class as in [14]. The membership values for the features are defined and these are encoded into the network as its connection weights. The hidden nodes of the network correspond to the intervals.

An interval $[F_{j1}, F_{j2}]$ is denoted as the range of feature F_j, covered by class C_k. The membership value of the interval is represented as $\mu([F_{j1}, F_{j2}]) = \mu$(between F_{j1} and F_{j2}). It is computed as

$$\mu(\text{between } F_{j1} \text{ and } F_{j2}) = \{\mu(\text{greater than } F_{j1}) * \mu(\text{less than } F_{j1})\}^{1/2}, \quad (2.6)$$

where

$$\mu(\text{greater than } F_{j1}) = \begin{cases} \{\mu(F_{j1})\}^{1/2}, & \text{if } F_{j1} \leq c_{prop}, \\ \{\mu(F_{j1})\}^2, & \text{otherwise}, \end{cases} \tag{2.7}$$

and

$$\mu(\text{less than } F_{j2}) = \begin{cases} \{\mu(F_{j2})\}^{1/2}, & \text{if } F_{j2} \geq c_{prop}, \\ \{\mu(F_{j2})\}^2, & \text{otherwise}, \end{cases} \tag{2.8}$$

Here c_{prop} denotes c_{jl}, c_{jm} and c_{jh} for each of the corresponding three overlapping fuzzy sets *low*, *medium*, and *high*. These are defined as in [15]. The output membership for the corresponding class C_k is obtained as in [10].

The interval which does not include the class is considered as complement of the interval. The complement of the interval $[F_{j1}, F_{j2}]$ of the feature F_j is the region where the class C_k does not lie and is defined as $[F_{j1}, F_{j2}]^c$. The linguistic membership values for $[F_{j1}, F_{j2}]^c$ are defined by $\mu([F_{j1}, F_{j2}]^c) = \mu(\text{not between } F_{j1} \text{ and } F_{j2})$. It is calculated as

$$\mu(\text{not between } F_{j1} \text{ and } F_{j2}) = max\{\mu(\text{less than } F_{j1}), \mu(\text{greater than } F_{j2})\} \tag{2.9}$$

since, not between F_{j1} and $F_{j2} \equiv$ less than F_{j1} OR greater than F_{j2}. The linguistic membership values for class C_k in the interval $[F_{j1}, F_{j2}]$ are denoted by $\{\mu_L([F_{j1}, F_{j2}]), \mu_M([F_{j1}, F_{j2}]), \mu_H([F_{j1}, F_{j2}])\}$. Similarly, for the complement of the interval, using Eq. 2.9, we have $\{\mu_L([F_{j1}, F_{j2}]^c), \mu_M([F_{j1}, F_{j2}]^c), \mu_H([F_{j1}, F_{j2}]^c)\}$.

A fuzzy multi-layer perceptron (FMLP) [15] with one hidden layer is considered, taking two hidden nodes corresponding to $[F_{j1}, F_{j2}]$ and its complement, respectively. Links are defined between the input nodes and the two nodes in the hidden layer iff $\mu_A([F_{j1}, F_{j2}])$ or $\mu_A([F_{j1}, F_{j2}]^c) \geq 0.5$, $\forall j$, where $A \in \{L, M, H\}$. The weight $w_{k_{\alpha_p}}^{(0)} {}_{jm}$ between the k_{α_p} node of the hidden layer (the hidden node corresponding to the interval $[F_{j1}, F_{j2}]$ for class C_k) and j_m ($m \in \{\text{first}(L), \text{second}(M), \text{third}(H)\}$)th node of the input layer corresponding to feature F_j is set by

$$w_{k_{\alpha_p}}^{(0)} {}_{jm} = p_k + \epsilon, \tag{2.10}$$

p_k is the a priori probability of class C_k and ϵ is a small random number. This hidden node is designated as positive node. A second hidden node k_{α_n} is considered for the compliment case and is termed as a negative node. Its connection weights are initialized as

$$w_{k_{\alpha_n}}^{(0)} {}_{jm} = (1 - p_k) + \epsilon. \tag{2.11}$$

It is to be mentioned that the method described above can suitably handle convex pattern classes only. In case of concave classes one needs to consider multiple intervals for a feature F_j corresponding to the various convex partitions that may be generated to approximate the given concave decision region. This also holds for the complement of the region in F_j in which a particular class C_k is not included.

Hence, in such cases, hidden nodes, positive and negative, are introduced for each of the intervals with connections being established by Eqs. 2.10 and 2.11 for the cases of a class belonging and not belonging to a region, respectively. This would result in multiple hidden nodes for each of the two cases. In this connection, one may note that a concave class may also be subdivided into several convex regions as in [9].

Let there be $(k_{pos} + k_{neg})$ hidden nodes, where $k_{pos} = \sum_{\alpha_p} k_{\alpha_p}$ and $k_{neg} = \sum_{\alpha_n} k_{\alpha_n}$ generated for class C_k such that $k_{pos} \geq 1$ and $k_{neg} \geq 1$. Now connections are established between kth output node (for class C_k) and only the corresponding $(k_{pos} + k_{neg})$ hidden nodes.

The connection weight $w_{kk_\alpha}^{(1)}$ between the kth output node and the k_αth hidden node is calculated from a series of equations generated as follows. For an interval as input for class C_k, the expression for output $y_k^{(2)}$ of the kth output node is given by

$$y_k^{(2)} = f(y_{k_\alpha}^{(1)} w_{kk_\alpha}^{(1)} + \sum_{r \neq \alpha} 0.5 w_{kk_r}^{(1)}) \tag{2.12}$$

where $f(.)$ is the sigmoid function and the hidden nodes k_r correspond to the intervals not represented by the convex partition α. Thus for a particular class C_k we have as many equations as the number of intervals (including *not*) used for approximating any concave and/or convex decision region C_k. Thereby, we can uniquely compute each of the connection weights $w_{kk_\alpha}^{(1)} \forall \alpha$ (corresponding to each hidden node k_α and class C_k pair).

The network architecture, so encoded, is then refined by training it on the pattern set supplied as input. In case of "all connections" between input and hidden layers, all the link weights are trained. In case of "selected connections" only the selected link weights are trained, while the other connections are kept clamped at zero. If the network achieves satisfactory performance, the classifier design is complete. Otherwise, node growing or link pruning is applied to the network as in [10].

2.5 Rough Fuzzy Multi-layer Perceptron

Rough fuzzy MLP (rfMLP) [1] is designed by associating rough sets with fuzzy MLP (FMLP). Rough sets are used to extract domain knowledge about data. The process of extracting the knowledge about data and encoding it into FMLP is described as follows:

Figure 2.1 shows a block diagram for knowledge encoding procedure for rough fuzzy MLP in which the data is first transformed into a 3-dimensional linguistic space. A threshold Th $(0.5 \leq Th < 1)$ is imposed on the resultant linguistic data in order to make the data into binary (0 or 1). The binary valued data is represented in a decision table $S = <U, A>$ with C, a set of condition attributes, and $D = \{d_1,..., d_n\}$, a set of decision attributes.

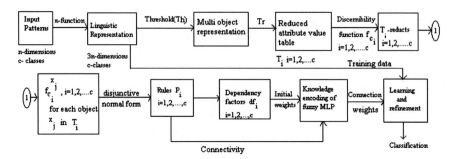

Fig. 2.1 Block diagram of the knowledge encoding procedure in rough fuzzy MLP [1]

(1) *Rule generation*: The decision table $S = <U, A>$ is divided into n tables S_i = $<U_i, A_i>$, $i = 1, 2, \ldots, n$ corresponding to n decision attributes $d_1, \ldots, d_n$, where $U = U_1 \bigcup U_2, \ldots, \bigcup U_n$ and $A_i = C \bigcup \{d_i\}$. The size of each S_i, $i = 1, 2, \ldots, n$, is first reduced with the help of a threshold on the number of occurrences of the same pattern of attribute values. The reduced decision table is denoted by τ_i, and $\{x_{i_1}, \ldots, x_{i_p}\}$ is a set of objects of U_i that occur in τ_i, $i = 1, 2, \ldots, n$. The discernibility matrix and discernibility function $f_{d_i}^{x_j}$ for every object $x_j \in \{x_{i_1}, \ldots, x_{i_p}\}$ are defined as in [1]. Then, $f_{d_i}^{x_j}$ is brought to its CNF a dependency rule r_i, viz., $P_i \rightarrow d_i$, where P_i is the disjunctive normal form (DNF) of $f_{d_i}^{x_j}$, $j \in \{i_1, \ldots, i_p\}$.

(2) *Knowledge encoding*: A $n_k \times 3n$-dimensional attribute value table for each class C_k is generated, where n_k indicates the number of objects in C_k. There are m sets $O_1, O_2, \ldots, O_m$ of objects in the table having identical attribute values and card(O_i) = n_{k_i}, $i = 1, \ldots, m$, such that $n_{k_1} \geq \cdots \geq n_{k_m}$ and $\sum_{j=1}^{m} n_{k_i} = n_k$. The attribute value table can now be represented as $m \times 3n$ array. Let $n'_{k_1}, \ldots, n'_{k_m}$ denote the distinct elements among $n_{k_1}, \ldots, n_{k_m}$ such that $n'_{k_1} \geq \cdots \geq n'_{k_m}$. A heuristic threshold function is defined as

$$Tr = \left\lceil \frac{\sum_{i=1}^{m} \frac{1}{n'_{k_i} - n'_{k_{(i+1)}}}}{TH} \right\rceil. \tag{2.13}$$

All entries having frequency less than Tr are eliminated from the table, resulting in the reduced attribute-value table. The main motive of introducing this threshold function lies in reducing the size of the resulting network thereby; eliminating noisy pattern representatives (having lower values of n_{k_i}) from the reduced attribute-value table. Based on the reduced attribute-value table, rough rules are generated as explained above. These are used to determine the dependency factors df_l, $l = 1, 2, \ldots, c$ which are then considered as the initial connection weights of the fuzzy MLP.

2.6 Architecture of Fuzzy Rough Granular Neural Networks

The fuzzy rough granular neural networks, FRGNN1 [4] and FRGNN2 [5] for classification are formulated by integrating fuzzy sets, fuzzy rough sets and multilayer perceptron. The advantages of the granular neural network include tractability, robustness, and close resemblance with human-like (natural) decision making for pattern recognition in ambiguous situations. It may be noted that while multilayer perceptron is a capable of classifying all types of real life data (both convex and non convex shapes), rough set handles uncertainty in class information, and fuzzy sets are used to resolve the overlapping regions among the classes. The information granules derived from the lower approximation of rough set are employed in the process of designing the network models. Each of the networks contains the input layer, the hidden layer and the output layer. A gradient decent method is used for training the networks. The number of nodes in input layer is determined by $3n$ dimensional granular input vector where n is total number of features in data. Every feature is transformed into 3-dimensional features, that constitute a granular input vector, along a feature axis of the three linguistic terms *low, medium* or *high*. The number of nodes in the hidden layer of FRGNN1 is set equal to the number of decision tables (c) where every decision table contains $3n$ dimensional linguistic training data belonging to the c classes. In FRGNN2, the number of nodes in the hidden layer is set is equal to c where c is number of classes, and only one decision table that consists the entire training data is used. The initial connection weights of FRGNN1 and FRGNN2, based on the decision tables, are defined using the concepts of fuzzy rough sets.

Let us now describe in detail the architecture of fuzzy rough granular neural networks (FRGNN1 [4] and FRGNN2 [5]) for classification. The networks use multi-layer perceptron with back propagation algorithm based on the gradient decent method for training purpose. The initial architecture of FRGNN1 and FRGNN2 is shown in Fig. 2.2. The FRGNN1 and FRGNN2 contain the input, hidden and output layers. The number of nodes/neurons in the input layer of FRGNN1 and FRGNN2 is set equal to 3-dimensional (3D) granular vector. While the number of nodes in the hidden layer of FRGNN1 is set equal to the number decision tables, this number is equal to the number of classes for FRGNN2. The number of nodes in the output layer for both FRGNN1 and FRGNN2 is the number of classes present in the data. Each neuron in the hidden layer is fully connected to the neurons in the previous layer and the next layer. The nodes in the input layer, non-computational nodes, receives an input and distributes to all the neurons in the next layer via its connection weights, during training. The input vector presented at the nodes in the input layer is in the granular form. The outputs of the input neurons are considered as the activation values to the nodes in the hidden layer. Similarly, the nodes in the output layer are received the outputs of the hidden layer nodes as their activation values. The node in the output layer corresponding to a class is assigned with a membership value for a pattern, while the other nodes, representing the other classes, are assigned with zeros. The fuzzy rough rules are initialized as initial connection weights of the links connecting

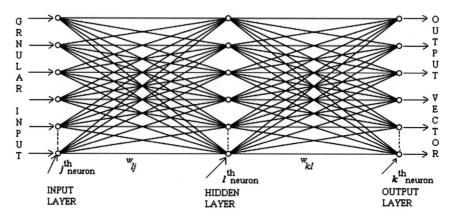

Fig. 2.2 Initial architecture of FRGNN1 and FRGNN2 with granular input layer, hidden layer and output layer with the same number of nodes

the nodes between input & hidden and hidden & output layers. The FRGNN1 and FRGNN2 are trained using back propagation algorithm with gradient-decent method which is described as follows:

Input

    D: Data set with training patterns in the granular form and their associated target vectors in terms of membership values and zeros (see Sect. 2.7.3 for details)

    α: Learning rate

    η: Momentum term

    $b_l = b$, Bias term which is kept constant at the node (l) in hidden and output layers

    Network: Granular feed-forward network

Method

1. Assign initial weights, in terms of dependency factors, between the nodes (units) in all the layers of the network, where the weights are determined by fuzzy rough sets based on a fuzzy similarity relation or fuzzy reflexive relation;

2. While the network is trained for all training patterns for a particular number of iterations {
   Forward Propagation:

3. For each unit j of the input layer, the output, say x_j, of an input unit, say I_j is

   {

   $x_j = I_j$;

   }

4. For each node l of the hidden or output layers, compute the output o_l of each unit l with respect to the previous layer, j, as

{

$$I_l = \sum_j w_{lj} x_j + b;$$

}

5. Apply logistic activation function to compute the output of each node l

{

$$\phi(x_l) = \frac{1}{1+e^{-I_l}};$$

} /* Note that, the logistic activation function is set at each node l in the hidden and the output layers*/

Back propagation

6. For each node in the output layer say k, compute the error using

{

$$Error_k = \phi(x_k)(1 - \phi(x_k))(TG_k - \phi(x_k));$$

}, where TG_k denotes the target value for each node k in the output layer.

7. Compute the error for lth node in the hidden layer by using the error obtained at kth node in the output layer (with respect to the next layer)

{

$$\Upsilon_l = \phi(x_l)(1 - \phi(x_l)) \sum_k Error_k w_{kl};$$

}, where $Error_k$ is an error value at the node k in the output layer (next layer).

8. Update the weight w_{lj} in the network using

{

$$\Delta w_{lj} = (\alpha) x_j \Upsilon_l;$$
$$\Delta w_{lj}(e) = (\alpha) x_j \Upsilon_l + (\eta) \Delta w_{lj}(e - 1);$$

} where Υ_l is the error value at the node l of the hidden or output layers.

9. For each constant value of bias b at node l of the hidden or output layers in the network

{

$$\Delta b = (\alpha) \Upsilon_l;$$
$$b(e) + = \Delta b(e - 1);$$

}

} /* end while loop*/

Output: Classified patterns

Here, the momentum term η is used to escape the local minima in the weight space. The value of the bias b is kept constant at each node of the hidden and output layers of the network, and e denotes the number of iterations. For every iteration, the training data is presented to the network for updating the weight parameters w_{lj} and bias b. The resulting trained network is used for classifying the test patterns.

2.7   Input Vector Representation

A formal definition of fuzzy granule is defined by the generalized constraint form "X isr R" where 'R' is a constrained relation, 'r' is a random set constraint which is a combination of probabilistic and posibilistic constraints, and 'X' is a fuzzy set which contains the values of the patterns in terms of linguistic terms *low*, *medium* and *high*. By using fuzzy-set theoretic techniques [12], a pattern x is assigned a membership value, based on a π function, to fuzzy set A and is defined as

$$A = \{(\mu_A(x), x)\}, \quad \mu_A(x) \in [0, 1] \tag{2.14}$$

where $\mu_A(x)$ represents the membership value of the pattern x. The π membership function, with range [0, 1] and $x \in \mathbf{R}^n$, is defined as

$$\pi(x, C, \lambda) = \begin{cases} 2(1 - \frac{\|x - C\|_2}{\lambda})^2, & \text{for } \frac{\lambda}{2} \leq \| x - C \|_2 \leq \lambda, \\ 1 - 2(\frac{\|x - C\|_2}{\lambda})^2, & \text{for } 0 \leq \| x - C \|_2 \leq \frac{\lambda}{2}, \\ 0, & \text{otherwise} \end{cases} \tag{2.15}$$

where $\lambda > 0$ is a scaling factor (radius) of the π function with C as a central point, and $\| \cdot \|_2$ denotes the Euclidean norm.

2.7.1   Incorporation of Granular Concept

An n-dimensional ith pattern is represented by a $3n$-dimensional linguistic vector [15] as

$$\vec{F}_i = [\mu_{low(F_{i1})}(\vec{F}_i), \mu_{medium(F_{i1})}(\vec{F}_i), \mu_{high(F_{i1})}(\vec{F}_i), \ldots, \mu_{high(F_{in})}(\vec{F}_i)] \tag{2.16}$$

where $\vec{F}_i$ is a feature vector for ith pattern, μ indicates the value of π- function corresponding to linguistic terms (fuzzy granule) *low*, *medium* or *high* along each feature axis. For example, the fuzzy granule for the feature F_{i1} of ith pattern (in terms of *low*, *medium* and *high*) is quantified as

$$F_{i1} \equiv \left\{ \frac{0.689}{L}, \frac{0.980}{M}, \frac{0.998}{H} \right\}.$$

When the input is numerical, we use the π-membership function (Eq. 2.15) with appropriate values of center C and scaling factor λ. Their selection is explained in the following section.

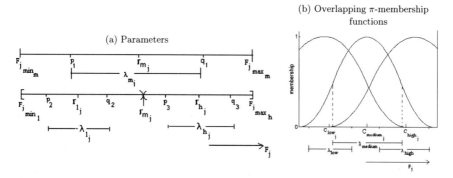

Fig. 2.3 Parameters for overlapping linguistic π-membership functions corresponding to linguistic terms, *low*, *medium* and *high* along the jth feature F_j

2.7.2 Choice of Parameters of π Membership Functions

Let $\{F_{ij}\}$ for $i = 1, 2, \ldots, s$, $j = 1, 2, \ldots, n$, represent a set of s patterns with n features, and $F_{j_{min_m}}$ and $F_{j_{max_m}}$ denote the minimum and maximum values along jth feature considering all the s patterns. Initially, the average of feature values of all the s patterns along the jth feature F_j is considered as the center of the linguistic term *medium* along that feature and denoted by r_{m_j}, as shown in Fig. 2.3a. Then, the average values (along the jth feature F_j) of the patterns having label values in the ranges $[F_{j_{min_m}}, r_{m_j})$ and $(r_{m_j}, F_{j_{max_m}}]$ are defined as the means of linguistic terms *low* and *high*, denoted by r_{l_j} and r_{h_j}, respectively. Similarly, considering the patterns with the values in the ranges $[F_{j_{min_m}}, r_{m_j})$ and $(r_{m_j}, F_{j_{max_m}}]$ along jth axis, we define $F_{j_{min_l}} = F_{j_{min_m}}$, $F_{j_{max_l}} = r_{m_j}$, $F_{j_{min_h}} = r_{m_j}$, and $F_{j_{max_h}} = F_{j_{max_m}}$. Then the center C and corresponding scaling factor λ for linguistic terms *low*, *medium* and *high* along the jth feature F_j are as follows:

$$
\begin{aligned}
C_{medium_j} &= r_{m_j}, \\
p_1 &= C_{medium_j} - \frac{F_{j_{max_m}} - F_{j_{min_m}}}{2}, \\
q_1 &= C_{medium_j} + \frac{F_{j_{max_m}} - F_{j_{min_m}}}{2}, \\
\lambda_{m_j} &= q_1 - p_1, \\
\lambda_{medium} &= \frac{\sum_{j=1}^{n} \lambda_{m_j}}{n}.
\end{aligned} \tag{2.17}
$$

$$
\begin{aligned}
C_{low_j} &= r_{l_j}, \\
p_2 &= C_{low_j} - \frac{F_{j_{max_l}} - F_{j_{min_l}}}{2}, \\
q_2 &= C_{low_j} + \frac{F_{j_{max_l}} - F_{j_{min_l}}}{2}, \\
\lambda_{l_j} &= q_2 - p_2, \\
\lambda_{low} &= \frac{\sum_{j=1}^{n} \lambda_{l_j}}{n}.
\end{aligned} \tag{2.18}
$$

$$
\begin{aligned}
C_{high_j} &= r_{h_j}, \\
p_3 &= C_{high_j} - \frac{F_{jmax_h} - F_{jmin_h}}{2}, \\
q_3 &= C_{highj} + \frac{F_{jmax_h} - F_{jmin_h}}{2}, \\
\lambda_{h_j} &= q_3 - p_3, \\
\lambda_{high} &= \frac{\sum_{j=1}^{n} \lambda_{h_j}}{n}.
\end{aligned}
\tag{2.19}
$$

The value of scaling factor for linguistic term *low* along the jth feature axis (λ_{l_j}) is defined as in Eq. 2.17, where the minimum and maximum values of λ_{l_j} are represented by p_1 and q_1. Moreover, the values of scaling factors for linguistic terms *medium* and *high* along the jth feature axis are defined as in Eqs. 2.18 and 2.19, where the corresponding minimum & maximum values are denoted by (p_2 & q_2) and (p_3 & q_3). The λ-value for a linguistic term is defined by taking the average of values of the linguistic term along all feature axes. Clearly, the λ-value for a particular linguistic term along all the axes would remain the same. Equations 2.17–2.19 automatically ensure that each input feature value along jth axis for a pattern $\overrightarrow{F}_i$ is assigned three membership values corresponding to the three-dimensional (3D) granular space of Eq. 2.16 in such a way that at least one of the $\mu_{low(F_{i1})}(\overrightarrow{F}_i)$, $\mu_{medium(F_{i1})}(\overrightarrow{F}_i)$ and $\mu_{high(F_{i1})}(\overrightarrow{F}_i)$ is greater than 0.5. In other words, this allows a pattern $\overrightarrow{F}_i$ to have a strong membership to at least one of the linguistic properties *low*, *medium* or *high*. This representation is shown diagrammatically in Fig. 2.3b.

2.7.3   Defining Class Membership at Output Node

The membership of ith pattern to kth class is defined as [12]

$$
\mu_k(\overrightarrow{F}_i) = \frac{1}{1 + (\frac{Z_{ik}}{f_d})^{f_e}}
\tag{2.20}
$$

where Z_{ik} is a weighted distance, and f_d and f_e are the denominational and exponential fuzzy generators controlling the amount of fuzziness in the class membership. Obviously, the class membership lies in [0, 1]. Here, the weighted distance Z_{ik} is defined in [15] as

$$
Z_{ik} = \sqrt{\sum_{j=1}^{n} \left[\sum_{p=1}^{3} \frac{1}{3} (\mu_p(F_{ij}) - \mu_p(o_{kj}))^2 \right]}, \quad \text{for } k = 1, 2, \ldots, c
\tag{2.21}
$$

where o_{kj} is the center of the jth feature vector from the kth class, μ_p is a membership value of the pth linguistic term (granule), and c is the number of classes. In the fuzziest case, we may use fuzzy modifier, namely, contrast intensification (INT) operator, to enhance the contrast in membership values of each pattern within that class in order to decrease the ambiguity in taking a decision.

2.7.4 Applying the Membership Concept to the Target Vector

The target vector at the output layer is defined in Eq. 2.22 by a membership value and zeros. For example, if a training pattern belongs to kth class, its desired output vector would have only one non-zero membership value corresponding to the kth node representing that class and zero value for the remaining nodes in the output layer. Therefore, for the ith training pattern $\overrightarrow{F_i}$ from the kth class, we define the desired output of the kth output node TG_k considering the fuzziest case, as

$$TG_k = \begin{cases} \mu_{INT}(\overrightarrow{F_i}), & \text{if the } i\text{th pattern is from } k\text{th class denoting} \\ & \text{the } k\text{th output node,} \\ 0, & \text{otherwise.} \end{cases} \qquad (2.22)$$

The network then back-propagates the errors with respect to the desired membership values at the output stage.

2.8 Fuzzy Rough Sets: Granulations and Approximations

In the context of fuzzy rough set theory, fuzzy set theory [8, 13] allows that an object belongs to a set and a couple of objects belong to a relation are assigned with membership values, and rough set theory allows the objects in a set that are assigned with the membership values belonging to the lower and approximations. Recall that Eq. 2.14 represents fuzzy set in U. A couple of objects x and y in U, the relation R is a mapping $U \times U \rightarrow [0, 1]$ such that $R = \{((x, y), R(x, y)) | R(x, y) \in [0, 1]\}$. For each $y \in U$, the R-foreset of y is the fuzzy set, say R_y, and is defined as $R_y(x) = R(x, y)$, for all x in U. The similarity between the two objects is represented by a fuzzy similarity relation and it should satisfy the following properties:

$$R(x, x) = 1 \quad \text{(reflexive)},$$
$$R(x, y) = R(y, x) \quad \text{(symmetry), and}$$
$$T(R(x, y)R(y, z)) \leq R(x, z) \quad \text{(T− transitivity)},$$

for all x, y and z in U. Given a T-norm, if R does not satisfy symmetry then R is called a fuzzy similarity relation (fuzzy tolerance relation), and if R does not satisfy symmetry and T-transitivity properties then R is called a fuzzy reflexive relation (fuzzy t-equivalence relation). Usually, the fuzzy T-equivalence relations constitute fuzzy T-equivalence class (fuzzy equivalence granule). The fuzzy T-equivalence classes are exploited in approximating the sets (concepts) in U. In fuzzy rough set theory, the following fuzzy logical connectives [8] are typically used in generalization of the lower and upper approximations of a set. For all x and $y \in [0, 1]$, an operator T mapping from $[0, 1]^2$ to $[0, 1]$ satisfies $T(1, x) = x$. We use T_M and T_L to represent T-norms and these are defined as

$$T_M(x, y) = min(x, y), \qquad\qquad (2.23)$$

$$T_L(x, y) = max(0, x + y - 1) \quad \text{(Lukasiewicz } t\text{-norm)}. \qquad (2.24)$$

for all x and $y \in [0, 1]$. On the other, a mapping $I : [0, 1]^2 \rightarrow [0, 1]$ such that $I(0, 0) = 1$, $I(1, x) = x$ for all $x \in [0, 1]$ where I is an implicator. For all x and $y \in [0, 1]$, the implicators I_M and I_L are defined as

$$I_M(x, y) = max(1 - x, y), \qquad\qquad (2.25)$$

$$I_L(x, y) = min(1, 1 - x + y) \quad \text{(Lukasiewicz implicator)}. \qquad (2.26)$$

2.8.1   Concepts of Fuzzy Rough Sets: Crisp and Fuzzy Ways

In fuzzy rough sets, an information system is a pair $(U, \mathcal{A})$ where $U = \{x_1, x_2, \ldots, x_s\}$ and $\mathcal{A} = \{a_1, a_2, \ldots, a_n\}$ are finite non empty sets of objects and conditional attributes, respectively. The values of conditional attributes can be quantitative (real valued). Let a be an attribute in $\mathcal{A}$. The fuzzy similarity relation R_a between any two objects x and y in U with respect to the attribute a is defined as [7]

$$R_a(x, y) = max \left(min \left(\frac{a(y) - a(x) + \sigma_a}{\sigma_a}, \frac{a(x) - a(y) + \sigma_a}{\sigma_a} \right), 0 \right) \qquad (2.27)$$

where σ_a denotes the standard deviation of the attribute a. It may be noted that the relation R_a does not necessarily satisfy the T-transitivity property.

A decision system $(U, \mathcal{A} \cup \{d\})$ is one kind of information system in which d ($d \notin \mathcal{A}$) is called a decision attribute and can be qualitative (discrete-valued). Based on these values, the set U is partitioned into non-overlapping sets/concepts corresponding to the decision concepts/decision classes, say $d_k, k = 1, 2, \ldots, c$. Each decision class can be represented by a crisp set or a fuzzy set. The belongingness of a pattern to each of the classes can be represented by decision feature containing membership values for all the patterns in the class. For a qualitative attribute a in $\{d\}$, the decision classes are defined in the following two methods.

2.8.1.1   Method I: Crisp Way of Defining Decision Classes (Crisp Case)

$$R_a(x, y) = \begin{cases} 1, & \text{if } a(x) = a(y), \\ 0, & \text{otherwise,} \end{cases} \qquad (2.28)$$

for all x and y in U. The crisp-valued decision class implies that objects in the universe U corresponding to the decision class would take values only from the set $\{0, 1\}$. A fuzzy set $A \subseteq U$ can be approximated only by constructing the lower and upper approximations of A with respect to crisp decision classes.

In real life problems, the data is generally ill defined, with overlapping class boundaries. Each pattern (object) in the set $A \subseteq U$ may belong to more than one class. To model such data, we extend the concept of a crisp decision granule into a fuzzy decision granule by inclusion of the fuzzy concept to crisp decision granules. The fuzzy decision classes are defined as follows:

2.8.1.2   Method II: Fuzzy Way of Defining Decision Classes (Fuzzy Case)

Consider a c-class problem domain where we have c decision classes of a decision attribute in a decision system. Let the n-dimensional vectors O_{kj} and V_{kj}, $j = 1, 2, ..., n$, denote the mean and standard deviation of the data for the kth class in the decision system. The weighted distance of a pattern $\overrightarrow{F_i}$ from the kth class is defined [12] as

$$Z_{ik} = \sqrt{\sum_{j=1}^{n} \left[\frac{F_{ij} - O_{kj}}{V_{kj}} \right]^2}, \quad \text{for } k = 1, 2, \ldots, c, \qquad (2.29)$$

where F_{ij} is the value of the jth feature (attribute) of the ith pattern. The parameter $\frac{1}{V_{kj}}$ acts as the weighting coefficient, such that larger the value of V_{kj} is, the less is the importance of the jth feature in characterizing the kth class. Note that when the value of a feature for all the patterns in a class is the same, the standard deviation of those patterns along that feature will be zero. In that case, we consider $V_{kj} = 0.000001$ (a very small value, instead of zero for the sake of computation), so that the distance Z_{ik} becomes high and the membership value of ith pattern to kth class, denoted by $\mu_k(\overrightarrow{F_i})$, becomes low where the membership value is defined using Eq. 2.20.

It may be noted that when a pattern $\overrightarrow{F_i}$ has different membership values corresponding to c decision classes then its decision attribute becomes quantitative, i.e., each pattern with varying membership value. The quantitative decision attribute can be made qualitative (as in Method I, crisp case) by two ways, namely, (i) by computing the average of the membership values over all the patterns in kth class to its own class, and assigning it to each pattern $\overrightarrow{F_i}$ in its kth decision class, and (ii) by computing the average of the membership values over all the patterns in kth class to the other classes, and assigning it to each pattern $\overrightarrow{F_i}$ in other decision classes (other than the kth class). So the average membership value (qualitative) of all the patterns in the kth class to its own class is defined as

$$D_{kk} = \frac{\sum_{i=1}^{m_k} \mu_k(\overrightarrow{F}_i)}{|m_k|}, \text{ if } k = u, \tag{2.30}$$

and the average membership values of all the patterns in the kth class to other decision classes are defined as

$$D_{ku} = \frac{\sum_{i=1}^{m_k} \mu_u(\overrightarrow{F}_i)}{|m_k|}, \text{ if } k \neq u, \tag{2.31}$$

where $|m_k|$ indicates the number of patterns in the kth class and k and $u = 1, 2, \ldots, c$.

For a qualitative attribute $a \in \{d\}$, the fuzzy decision classes are defined as

$$R_a(x, y) = \begin{cases} D_{kk}, & \text{if } a(x) = a(y), \\ D_{ku}, & \text{otherwise,} \end{cases} \tag{2.32}$$

for all x and y in U. Here D_{kk} corresponds an average membership value of all the patterns that belong to the same class ($k = u$), and D_{ku} corresponds to the average membership values of all the patterns from the classes other than k ($k \neq u$).

For any $\mathcal{B} \subseteq \mathcal{A}$, the fuzzy $\mathcal{B}$-indiscernibility relation $R_\mathcal{B}$ based on the fuzzy similarity relation R is induced by

$$R_\mathcal{B}(x, y) = R_a(x, y), a \in \mathcal{B}, \tag{2.33}$$

The fuzzy lower and upper approximations of the set $A \subseteq U$, based on fuzzy similarity relation R, are defined as [16]

$$(R_\mathcal{B} \downarrow R_d)(y) = \inf_{x \in U} I(R_\mathcal{B}(x, y), R_d(x)), \tag{2.34}$$

$$(R_\mathcal{B} \uparrow R_d)(y) = \sup_{x \in U} T(R_\mathcal{B}(x, y), R_d(x)), \tag{2.35}$$

for all y in U. Here the fuzzy logic connectives, Lukasiewicz T-norm T and implicator I, are exploited in defining the lower and upper approximations. The fuzzy positive region, based on the fuzzy $\mathcal{B}$-indiscernibility relation, is defined in [16] as

$$POS_\mathcal{B}(y) = \left(\bigcup_{x \in U} R_\mathcal{B} \downarrow R_d x \right)(y), \tag{2.36}$$

for all x and $y \in U$. Here, $POS_\mathcal{B}(y)$ characterizes the positive region with a maximum membership value of pattern y. Considering y belonging to a set A which is a subset of U, the fuzzy positive region Eq. 2.37 is simplified as [2]

$$POS_\mathcal{B}(y) = (R_\mathcal{B} \downarrow R_d x)(y). \tag{2.37}$$

The dependency degree of a set of attributes $\mathcal{B} \subseteq \mathcal{A}$, denoted by $\gamma_\mathcal{B}$, is defined as

$$\gamma_\mathcal{B} = \frac{\sum_{x \in U} POS_\mathcal{B}(x)}{|U|} \tag{2.38}$$

where $| \cdot |$ denotes the cardinality of a set U, and γ is $0 \leq \gamma \leq 1$.

It may be noted that the aforesaid procedure of determining the dependency factor of each conditional attribute with respect to the decision classes (either crisp case or fuzzy case) enables the fuzzy rough granular neural networks to define their initial weight parameters from the training samples. The initial weights in the fuzzy case (Method II) are seen to provide better performance in handling the overlapping class boundaries than those of the crisp case (Method I). These are explained in the following sections.

2.9 Configuration of the Granular Neural Networks Using Fuzzy Rough Sets

In this section, we explain the concept of granulation by partitioning data in the decision tables, computing equivalence relations using the fuzzy similarity relation and approximation of a set using rough approximations. Based on the granulation principle, the initial weights of the FRGNN1 and FRGNN2 are then determined; thereby providing a knowledge based network. Such a network is found to be more efficient than fuzzy MLP [15] and other similar types of granular neural networks [1, 3] as shown in experiments. During training, the networks search for a set of connection weights that corresponds to some local minima. Note that there may be a large number of such minima corresponding to various good solutions. Therefore, if we can initially set weights of the network so as to correspond one such solution, the searching space may be reduced and learning thereby becomes faster. Further, the architecture of the network can be made simpler by fixing the number of nodes in the hidden layer based on the class information. These are the characteristics that the fuzzy rough granular neural networks are capable to achieve. At first, the knowledge encoding procedure of FRGNN1 is described using the concepts of fuzzy rough set and fuzzy similarity relation. Then, it is also described for FRGNN2. The procedure of knowledge encoding is explained in Fig. 2.4 using block diagram.

2.9.1 Knowledge Encoding Procedures

Method of FRGNN1: The knowledge encoding procedure for FRGNN1 [4] is explained as follows. We use the aforesaid decision table $S = (U, \mathcal{A} \cup \{d\})$ with $\mathcal{A} = \{a_1, a_2, \ldots, a_n\}$ its set of conditional attributes, and with decision attributes $\{d\}$

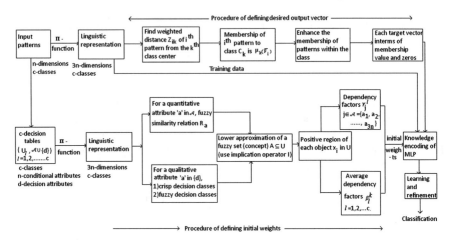

Fig. 2.4 Block diagram of knowledge encoding

where $U = \{x_1, x_2, \ldots, x_s\}$, its set of objects, form c classes and the objects having labeled values corresponding to each n-dimensional conditional attribute. We split the decision table $S = (U, \mathcal{A} \cup \{d\})$ into c decision tables $S_l = (U_l, \mathcal{A} \cup \{d\})$, $l = 1, 2, \ldots, c$, corresponding to c classes, and the objects are added to each decision table from all the c classes succession. Moreover S_l satisfies the following conditions:

$$(i)\, U_l \neq \emptyset, \quad (ii)\, \bigcup_{l}^{c} U_l = U, \quad (iii)\, \bigcap_{l=1}^{c} U_l = \emptyset. \tag{2.39}$$

The size of each $S_l, l = 1, 2, \ldots, c$, is dependent on the available number of objects from all the classes. If all the classes have an equal number of patterns, then the number of objects that will be added to each S_l will be the same; otherwise, it will be different.

Let us consider the case of feature F_1, when $j = 1$ (attribute a_1). The ith pattern $\vec{F}_i$ is mapped to a point in the 3D feature space of $\mu_{low(F_{i1})}(\vec{F}_i)$, $\mu_{medium(F_{i1})}(\vec{F}_i)$, $\mu_{high(F_{i1})}(\vec{F}_i)$ using Eq. 2.16. In this manner, an n-dimensional attribute valued decision table can be transformed into a $3n$-dimensional attribute valued decision table. We apply the following procedure for the decision table $S_l, l = 1, 2, \ldots, c$.

S1: Define granulation structures, using the fuzzy similarity relation in Eq. 2.27, for each conditional attribute by generating a fuzzy similarity matrix.

S2: Use the fuzzy similarity relations and compute lower approximations of Eq. 2.34 of each concept for each conditional attribute with respect to the decision classes using Eqs. 2.28 or 2.32.

S3: Calculate the fuzzy positive region, using Eq. 2.37, of an object in the concept for each conditional attribute.

S4: Calculate dependency degrees, using Eq. 2.38, for the conditional attributes, and the dependency degrees are initialized between the nodes of the input layer and the hidden layer of networks as connection weights.

S5: Calculate the average of the dependency degrees of all conditional attributes with respect to every decision class and define the average values between the nodes of the hidden layer and the output layer as initial weights.

Let us now design the initial structure of FRGNN1. It contains three layers. The number of nodes in the input layer is set equal to 3n-dimensional attributes and it is represented by c classes in the output layer. The hidden layer nodes are modeled with the class information. We explain the procedure for initializing the initial weights of FRGNN1.

Let γ_j^l denote the dependency degree of jth conditional attribute in lth decision table S_l where $j \in \mathcal{A} = \{a_1, a_2, ..., a_{3n}\}$ and $l = 1, 2, ..., c$. The weight w_{lj} between an input node j and a hidden node l (see Fig. 2.2) is defined as follows: For instance, the decision table S_1, when $l = 1$,

$$\gamma_j^1 = \frac{\sum_{x \in U} POS_j(x)}{|U|}. \tag{2.40}$$

Let β_l^k denote the average dependency degrees for all conditional attributes with respect to kth decision class d_k, $k = 1, 2, ..., c$, in lth decision table $S_1, l = 1, 2, ..., c$. The weight w_{kl} between the hidden node l and the output node k (see Fig. 2.2) is defined as follows: For instance, the decision table S_1, when $l = 1$, and its kth decision class d_k, $k = 1, 2, ..., c$,

$$\beta_1^k = \frac{\sum_{j=1}^{3n} \gamma_j^k}{|3n|}, \tag{2.41}$$

where γ_j^k is defined as

$$\gamma_j^k = \frac{\sum_{x \in A_k} POS_j(x)}{|A_k|}, \tag{2.42}$$

where $A_k \subseteq U$ represents kth set that contains objects with respect to kth decision class d_k and $POS_j(x)$ indicate the fuzzy positive membership value of an object x in A_k, corresponding to jth attribute.

Similar procedure is applied to the rest of the decision tables S_l ($l = 2, 3, ..., c$). Then the dependency degrees of conditional attributes, γ_j^l, for $l = 1, 2, ..., c$, are used as the initial weights between the nodes of the input layer and the hidden layer (see Step 4). The initial weights between the nodes of the hidden and the output layers are set as β_l^k (see Step 5). The connection weights, so encoded, are refined by training the network on a set of patterns supplied as input.

Method of FRGNN2: We use the aforesaid decision system $S = (U, \mathcal{A} \cup \{d\})$ where $U = \{x_1, x_2, ..., x_s\}$ represents a set of objects from c-classes, and $\mathcal{A} = \{a_1, a_2, ..., a_n\}$ & $\{d\}$ denote set of conditional attributes and decision attributes, respectively. The objects in universe U are classified into k number of non-overlapping sets/concepts, A_k, corresponding to decision classes, $d_k, k = 1, 2, \ldots, c$. The fuzzy reflexive relation R_a between two objects x and y in U corresponding to an attribute a is defined as [5]

$$
R_a(x, y) = \begin{cases} max \left(min \left(\frac{a(y)-a(x)+\sigma_{a_k}}{\sigma_{a_k}}, \frac{a(x)-a(y)+\sigma_{a_k}}{\sigma_{a_k}} \right), 0 \right), \\ \qquad\qquad if \quad a(x)\&a(y) \in A_k, \\ \\ max \left(min \left(\frac{a(y)-a(x)+\sigma_{a_u}}{\sigma_{a_u}}, \frac{a(x)-a(y)+\sigma_{a_u}}{\sigma_{a_u}} \right), 0 \right), \\ \qquad\qquad if \quad a(x) \in A_k, a(y) \in A_u, \\ \qquad\qquad\qquad and \quad k \neq u \end{cases} \tag{2.43}
$$

where k and $u = 1, 2, \ldots, c$, and σ_{a_k} and σ_{a_u} represent the standard deviation of sets (concepts) A_k and A_u, respectively, corresponding to an attribute a.

Let us consider the case of feature F_j (attribute a_j) in the decision table S. The ith representative pattern $\overrightarrow{F}_i$ is mapped to a point in the 3D granular space of $\mu_{low(F_{i1})}(\overrightarrow{F}_i)$, $\mu_{medium(F_{i1})}(\overrightarrow{F}_i)$ and $\mu_{high(F_{i1})}(\overrightarrow{F}_i)$ by using Eq. 2.16. In this manner an n-dimensional attribute valued decision table S is transformed into a $3n$-dimensional attribute valued decision table S. The following steps are applied on the decision table S to determine the initial connection weights of FRGNN2.

(S1) Develop the fuzzy reflexive relation matrix using Eq. 2.43, corresponding to every conditional attribute, where row in the matrix is a granulation structure.
(S2) Use Steps 2 and 3 in Sect. 2.9.1.
(S3) Calculate dependency degree using Eq. 2.38 for each conditional attribute with respect to the decision classes. The resulting values are determined as initial weights between nodes in the input layer and the hidden layer.
(S4) At first, calculate the average of all the dependency degrees of all the conditional attributes with respect to every decision class. Then, divide the resulting average values by the number of nodes of the output layer, representing the number of classes, and assign those values between the nodes of the hidden and the output layers of the network as its initial connection weights.

Now we explain the knowledge encoding procedure of FRGNN2. Let γ_j^l denote the dependency degree of jth conditional attribute in decision table S where the number of decision classes in the decision table is c. The weight w_{lj} between the input node j and the hidden node l (see Fig. 2.2) is defined as

$$
\gamma_j^l = \frac{\sum_{x \in A_l} POS_j(x)}{|A_k|}, \quad l = 1, 2, \ldots, c, \tag{2.44}
$$

where A_l represents lth concept corresponding to decision class d_l.

Let β_l denote the average dependency degree of all the conditional attributes with respect to lth decision class. It is computed as

$$\beta_l = \frac{\sum_{j=1}^{3n} \gamma_j^l}{|3n|}, \quad l = 1, 2, \ldots, c. \tag{2.45}$$

The weight w_{kl} between node l in hidden layer and node k in the output layer is defined as $\frac{\beta_l}{|l|}$.

2.9.2 Examples for Knowledge Encoding Procedure

We explain the knowledge encoding procedure of FRGNN1 and FRGNN2 in crisp case (Method I) and fuzzy case (Method II) using data [7] in Table 2.1.

Example of FRGNN1 in crisp case:

The knowledge encoding procedure of FRGNN1 in crisp case (Method I) is explained as follows. Each conditional attribute in Table 2.1 is transformed into 3D attributes corresponding to *low*, *medium* and *high*. The 3D conditional attributes and the crisp decision classes are shown in Table 2.2.

We apply step S1 of Sect. 2.9.1 to Table 2.2. The fuzzy similarity matrix for a conditional attribute L_1 is provided as an example.

Table 2.1 Dataset

U	a	b	c	d
x_1	−0.4	−0.3	−0.5	1
x_2	−0.4	0.2	−0.1	2
x_3	−0.3	−0.4	0.3	1
x_4	0.3	−0.3	0	2
x_5	0.2	−0.3	0	2
x_6	0.2	0	0	1

Table 2.2 Decision table with 3-dimensional conditional attributes and decision attribute in crisp case

U	L_1	M_1	H_1	L_2	M_2	H_2	L_3	M_3	H_3	d
x_1	0.875	0.3950	0	0.9296	0.9243	0	0.125	0.3472	0	1
x_2	0.875	0.3950	0	0	0.2608	0.125	0	0.9861	0.3828	2
x_3	0.5	0.6975	0	0.3828	0.7391	0	0.125	0.875	0	1
x_4	0	0.3024	0.5	0.9296	0.9243	0	0	0.875	0.9296	2
x_5	0	0.6049	0.875	0.9296	0.9243	0	0	0.875	0.9296	2
x_6	0	0.6049	0.875	0	0.8132	0.125	0	0.875	0.9296	1

$$R_{L_1}(x, y) = \begin{pmatrix} 1 & 1 & 0.555 & 0 & 0 & 0 \\ 1 & 1 & 0.555 & 0 & 0 & 0 \\ 0.555 & 0.555 & 1 & 0.407 & 0.407 & 0.407 \\ 0 & 0 & 0.407 & 1 & 1 & 1 \\ 0 & 0 & 0.407 & 1 & 1 & 1 \\ 0 & 0 & 0.407 & 1 & 1 & 1 \end{pmatrix}$$

Similarly, the similarity matrices for the attributes $R_{M_1}(x, y)$, $R_{H_1}(x, y)$, $R_{L_2}(x, y)$, $R_{M_2}(x, y)$, $R_{H_2}(x, y)$, $R_{L_3}(x, y)$, $R_{M_3}(x, y)$, $R_{H_3}(x, y)$ are also determined. We then calculate the lower approximations, using step S2, for the concepts $A_1 = \{1, 3, 6\}$ and $A_2 = \{2, 4, 5\}$ for every conditional attribute with respect to the decision classes d_1 and d_2 under decision attribute column d where y belongs to A_1 & A_2 and x belongs to U.

We now apply step S2 in Sect. 2.9.1 to the concept A_1,
$(R_{L_1} \downarrow R_d x)(y) = \inf_{x \in U} I\{R_{L_1}(x, y), R_d(x, y)\}$.
For object x_1, this is
$(R_{L_1} \downarrow R_d x)(x_1) = \inf_{x \in U} I\{R_{L_1}(x, 1), R_d(x, 1)\}$,
$\qquad = \min \{I(1, 1), I(1, 0), I(0.555, 1), I(0, 0), I(0, 0), I(0, 1)\}$,
$\qquad = 0.0$.
The lower approximation values for objects x_3 and x_6 are 0.444 and 0.0, respectively.

We also apply step S2 in Sect. 2.9.1 to the concept A_2,

$$(R_{L_1} \downarrow R_d x)(x_2) = 0.0, (R_{L_1} \downarrow R_d x)(x_4) = 0.0, (R_{L_1} \downarrow R_d x)(x_5) = 0.0.$$

The dependency degree of a conditional attribute, $\gamma_{\{L_1\}}$, is 0.074. The dependency degrees for the remaining conditional attributes are defined as

$$\begin{aligned} \gamma_{\{M_1\}} &= 0.031, & \gamma_{\{H_1\}} &= 0.074, \\ \gamma_{\{L_2\}} &= 0.077, & \gamma_{\{M_2\}} &= 0.506, \\ \gamma_{\{H_2\}} &= 0.000, & \gamma_{\{L_3\}} &= 0.042, \\ \gamma_{\{M_3\}} &= 0.111, \text{ and} & \gamma_{\{H_3\}} &= 0.233. \end{aligned}$$

These values are considered as the initial connection weights from nine input layer nodes to one hidden layer node of the FRGNN1 corresponding to class 1. We define the connection weights between the hidden layer node and the output layer nodes as follows:

For a qualitative attribute $a \in \{d\}$, the lower approximations are equivalent to positive degrees of objects in the concepts A_1 and A_2. The positive membership values of objects in the concept $A_1 = \{1, 3, 6\}$ corresponding to the attribute L_1 are

$$POS_{L_1}(x_1) = 0.0, \ POS_{L_1}(x_3) = 0.444, \ POS_{L_1}(x_6) = 0.0.$$

The dependency degree is

$$\gamma_{\{L_1\}}(A_1) = \frac{0.0 + 0.444 + 0.0}{x_3} = 0.148.$$

The positive membership values of objects for the attribute L_1 are

$$POS_{L_1}(x_2) = 0.0, \; POS_{L_1}(x_4) = 0, \; POS_{L_1}(x_5) = 0.0.$$

The dependency degree is

$$\gamma_{\{L_1\}}(A_2) = \frac{0.0 + 0.0 + 0.0}{3} = 0.0.$$

The dependency degrees for a typical attribute M_1, with respect to the concepts A_1 and A_2, are given as follows:

$$\gamma_{\{M_1\}}(A_1) = 0.031, \qquad \gamma_{\{M_1\}}(A_2) = 0.031.$$

The dependency values for the remaining attributes are also computed and the average dependency degrees of all the conditional attributes with respect to the decision classes are characterized by $\gamma(A_1) = 0.113$ and $\gamma(A_2) = 0.060$. The connection weights between the hidden layer node and two output layer nodes that correspond to classes 1 and 2 are initialized with 0.113 & 0.060, respectively.

Example of FRGNN1 in fuzzy case:

Table 2.3 provides 3D conditional attributes and fuzzy decision classes, obtained using Eqs. 2.30 and 2.31, under the decision attribute columns D_{kk} and D_{ku}, respectively.

We then calculate the lower approximations, using step S2, of the concepts $A_1 = \{x_1, x_3, x_6\}$ and $A_2 = \{x_2, x_4, x_5\}$ for every conditional attribute with respect to the fuzzy decision classes under the columns of fuzzy decision attributes D_{kk} and D_{ku} where y belongs to A_1 or A_2 and x belongs to U.

For the concept $A_1 = \{1, 3, 6\}$,

$$(R_{L_1} \downarrow R_d x)(y) = \inf_{x \in U} I\{R_{L_1}(x, y), R_d(x, y)\}.$$

Table 2.3 Decision table with 3D conditional attributes and decision attributes in fuzzy case

U	L_1	M_1	H_1	L_2	M_2	H_2	L_3	M_3	H_3	d	D_{kk}	D_{ku}
x_1	0.875	0.395	0	0.929	0.924	0	0.125	0.347	0	1	0.257	0.180
x_3	0.5	0.697	0	0.382	0.739	0	0.125	0.875	0	1	0.257	0.180
x_6	0	0.604	0.875	0	0.813	0.125	0	0.875	0.929	1	0.257	0.180
x_2	0.875	0.395	0	0	0.260	0.125	0	0.986	0.382	2	0.276	0.084
x_4	0	0.302	0.5	0.929	0.924	0	0	0.875	0.929	2	0.276	0.084
x_5	0	0.604	0.875	0.929	0.924	0	0	0.875	0.929	2	0.276	0.084

For object x_1, this is

$$(R_{L_1} \downarrow R_d x)(x_1) = \inf_{x \in U} I\{R_{L_1}(x, x_1), R_d(x, x_1)\},$$
$$= \min \{I(1, 0.257), I(1, 0.180), I(0.555, 0.257),$$
$$I(0, 0.180), I(0, 0.180), I(0, 0.257)\},$$
$$= 0.180.$$

The dependency values for object x_3 and x_6 are obtained as 0.257 and 0.180, respectively. For the concept $A_2 = \{x_2, x_4, x_5\}$,

$$(R_{L_1} \downarrow R_d x)(x_2) = 0.084, \ (R_{L_1} \downarrow R_d x)(x_4) = 0.084, \ (R_{L_1} \downarrow R_d x)(x_5) = 0.084.$$

The dependency degree of the conditional attribute L_1 is $\gamma_{\{L_1\}} = 0.145$. Similarly, the dependency degrees for the remaining attributes are calculated. The resulting dependency degrees are used as the initial connection weights between the nodes in input layer and the hidden layer node of the FRGNN1 that corresponds to class 1. Similarly, one can determine from Table 2.3 the initial connection weights between the hidden layer node and the output layer nodes of FRGNN1. The resulting average dependency degrees of all the conditional attributes with respect to the decision classes d_1 and d_2 are characterized by $\gamma(A_1) = 0.209$ and $\gamma(A_2) = 0.113$. These values are initialized between one node in hidden layer and two nodes in output layer that correspond to classes 1 and 2.

So far we have discussed the procedure for determining the connection weights corresponding to one decision table. If the similar procedure is applied to the other decision table in both crisp and fuzzy cases then a fully connected FRGNN1 with initial weight parameters would be generated in both the cases.

Example of FRGNN2 in crisp case:

We use the data in Table 2.1 to explain the knowledge encoding procedure of FRGNN2. The 3D conditional attributes and the crisp decision classes under decision attribute columns are shown in Table 2.2. Step S1 of Sect. 2.9.1 is applied to Table 2.2 to define the fuzzy reflexive relational matrices for every conditional attribute. The fuzzy reflexive relational matrix for the attribute L_1, $R_{L_1}(x, y)$ for x and y in U, is provided as an example.

$$R_{L_1}(x, y) = \begin{pmatrix} 1 & 0.938 & 0.184 & 1 & 0.194 & 0.194 \\ 0.938 & 1 & 0.246 & 0.938 & 0.256 & 0.256 \\ 0.184 & 0.246 & 1 & 0.194 & 1 & 1 \\ 1 & 0.938 & 0.184 & 1 & 0.194 & 0.194 \\ 0.184 & 0.246 & 1 & 0.194 & 1 & 1 \\ 0.184 & 0.246 & 1 & 0.194 & 1 & 1 \end{pmatrix}$$

Similarly, the fuzzy reflexive relational matrices $R_{M_1}(x, y)$, $R_{H_1}(x, y)$, $R_{L_2}(x, y)$, $R_{M_2}(x, y)$, $R_{H_2}(x, y)$, $R_{L_3}(x, y)$, $R_{M_3}(x, y)$, and $R_{H_3}(x, y)$ are also determined. The membership values of the patterns belonging to the lower approximations of the concepts $A_1 = \{x_1, x_3, x_6\}$ and $A_2 = \{x_2, x_4, x_5\}$ arecomputed using step S2 in Sect. 2.9.1.

Here, the lower approximations are equal to the positive membership values. The dependency values of the attributes, based on the positive membership values, with respect to the decision classes, are computed using step S3 in Sect. 2.9.1. The dependency values of attribute L_1 with respect to class1, $\gamma_{\{L_1\}}(A_1)$, and with respect to class2, $\gamma_{\{L_1\}}(A_2)$, are given as examples.

$$\gamma_{\{L_1\}}(A_1) = 0.106, \qquad \gamma_{\{L_1\}}(A_2) = 0.$$

The weights between nine input nodes that correspond to nine attributes (L_1, M_1, H_1, L_2, M_2, H_2, L_3, M_3 and H_3) and one hidden node are initialized with the dependency values with respect to class 1 (0.106, 0.030, 0.0, 0.107, 0.089, 0.0, 0.083, 0.175 and 0.239), and other hidden node are initialized with the dependency values with respect to class2 (0, 0.030, 0.106, 0.0, 0.158, 0.0, 0.0, 0.035 and 0.107).

The average values of the dependency values with respect to classes 1 & 2 are 0.092 & 0.048, respectively. The weights between two hidden nodes & one output node that corresponds to class 1 are initialized with 0.046 & 0.046 (= 0.092/2), and the other output node that corresponds to class 2 are initialized with 0.024 & 0.024 (0.048/2). The procedure for determining the connection weights of FRGNN2 in fuzzy case (Method II) are also determined in similar way to that of weights in crisp case (Method I).

2.10   Experimental Results

The FRGNN1 and FRGNN2 algorithms are implemented in C-language. The characteristics of several real life data sets are provided in Table 2.4. The first four data sets in the table are used for FRGNN1 and the rest in the table are used for FRGNN2. The data sets, except Telugu vowel, are collected from the UCI Machine Learning

Table 2.4 Characteristics of the data sets for FRGNN1 and FRGNN2

Dataset name	Num of patterns	Features	Classes
Data used for FRGNN1			
Telugu vowel	871	3	6
Sonar	208	60	2
Lymphography	148	18	4
Thyroid	215	5	3
Data used for FRGNN2			
Telugu vowel	871	3	6
Image segmentation	2310	19	7
Sonar	208	60	2
Spam base	4601	57	2

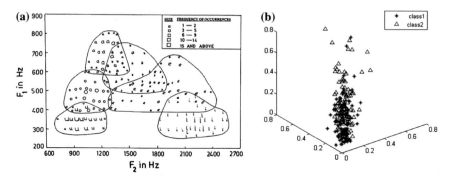

Fig. 2.5 a Vowel data in the F_1-F_2 plane. **b** Sonar data in the F_1-F_2-F_3 space

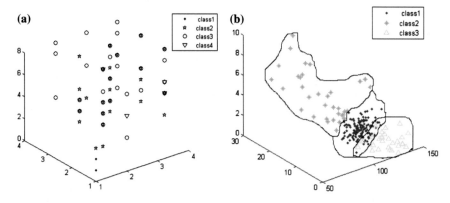

Fig. 2.6 a Lymphography data in the F_1-F_2-F_3 space. **b** Thyroid data in the F_1-F_2-F_3 space

Repository [11]. Figure 2.5a shows a two-dimensional (2D) projection of the 3D feature space of the six vowel classes in the F_1-F_2 plane. Figure 2.5b shows a 3D projection of the 60D feature space of two classes of sonar data in the F_1-F_2-F_3 space. Figure 2.6a shows a 3D projection of the 18D feature space of the four classes of lymphography data in the F_1-F_2-F_3 space. Figure 2.6b shows a 3D projection of the 5D feature space of the three classes of thyroid data in the F_1-F_2-F_3 space.

Results of FRGNN1

During learning of FRGNN1, we have used an v-fold cross validation design with stratified sampling. The training is done on $(v - 1)$ folds of the data selected randomly from each of the classes. The remaining one fold data is considered as the test set. This is repeated v times, and the overall performance of the FRGNN1 is computed taking an average over v sets of folds. In the experiment, we considered v = 10 or 9 based on the size of the data sets. The parameters in Eq. 2.20 were chosen as $f_d = 6$ and $f_e = 1$ for all the data sets. The momentum parameter α, learning rate η, and the bias b of the FRGNN1 traversed a range of values between 0 and 1, and had different values depending on the folds used for learning. For example, in case of the vowel data, the appropriate values were $\alpha = 0.98$, $\eta = 0.0858$, $b = 0.00958$ in crisp case, and $\alpha = 0.958$, $\eta = 0.06558$, $b = 0.0958$ in fuzzy case. It is observed that the FRGNN1 converges to a local minimum at the 1500th epoch for vowel data.

Before providing the performance of the FRGNN1 on the test sets for all the data, we explain the implementation process for its training on Telugu vowel data, as an example. Here, the FRGNN1 has nine nodes in the input layer, and six nodes in each of hidden and output layers. The data is first transformed into a 3D granular space using Eq. 2.16. The appropriate values of the center C and scaling factor λ for each feature of granules, e.g., *low*, *medium* or *high*, are determined as $C_{low_1} = 368.932$, $C_{medium_1} = 470.482$, $C_{high_1} = 583.616$, $C_{low_2} = 1110.323$, $C_{medium_2} = 1514.684$, $C_{high_2} = 2047.021$, $C_{low_3} = 2359.322$, $C_{medium_3} = 2561.021$, $C_{high_3} = 2755.891$, and $\lambda_{low} = 586.666$, $\lambda_{medium} = 1300.000$, $\lambda_{high} = 666.666$. Then, the 3-dimensional patterns have numerical components which are mapped into a 9D granular space with components L_1, M_1, H_1, L_2, M_2, H_2, L_3, M_3, H_3, while the desired vector has components d_1, d_2, d_3, d_4, d_5, d_6 corresponding to the six vowel classes. Table 2.5 provides some examples of a granular input vector $\overrightarrow{F}_i$, and the desired output vector d_k, $k = 1, 2, \ldots, 6$, for a set of sample patterns.

Knowledge extraction procedure for Telugu vowel data:

Let the decision table $S = (U, \mathcal{A} \cup \{d\})$ be used to represent the entire training data set. The decision table S is then divided into six decision tables corresponding to six vowel classes, namely, $S_l = (U_l, \mathcal{A} \cup \{d\})$. Every S_l, $l = 1, 2, \ldots, 6$, need not be of the same size. Each S_l is transformed into 3D granular space using Eq. 2.5. We apply the knowledge encoding procedure (which was explained before to the data in Table 2.1) for each decision table S_l, and the resulting domain knowledge is then encoded into the FRGNN1 in the form of initial connection weights. Table 2.6 provides the initial connection weights of the FRGNN1 in fuzzy case (see Method II of Sect. 2.8). Values of the fuzzifiers f_d and f_e in Eq. 2.20 were considered to be 1. The complete FRGNN1 architecture thus generated for the vowel data is shown in Fig. 2.7. The network is trained for refining the initial weight parameters in the presence of training samples. The trained network is then used to classify the set of test patterns.

Note that, if the number of training patterns in any class is less than the number of classes, then the average dependency factors of all the conditional attributes with respect to a decision class may not possible to define. To avoid this, we include some training patterns with zero attribute value to that particular class so as to make the number of training patterns in that class equal to the number of classes.

Performance of the FRGNN1 on Test Sets:

The experimental results of the FRGNN1 for Telugu vowel, sonar, lymphography and thyroid are shown in Table 2.7. Results correspond to three types of the initial weights of the FRGNN1. These are (i) random numbers in [-0.5, 0.5], (ii) crisp case (Method 1 of Sect. 2.8) and (iii) fuzzy case (Method 2 of Sect. 2.8). One may note that considering the random initial weights (i.e., type (i)) makes the FRGNN1 equivalent to a fuzzy MLP [15].

Table 2.5 Examples of input and target vectors for FRGNN1. A pattern with three input features is mapped into 3×3D granular space with components L_1, M_1, H_1, L_2, M_2, H_2, L_3, M_3 and H_3, while the desired vector has components d_1, d_2, d_3, d_4, d_5, d_6 corresponding to the 6 vowel classes

| Input features | | | Input vector | | | | | | | | | Desired vector | | | | | |
F_1	F_2	F_3	L_1	M_1	H_1	L_2	M_2	H_2	L_3	M_3	H_3	d_1	d_2	d_3	d_4	d_5	d_6
700	1500	2600	0.37	0.94	0.93	0.22	0.99	0.06	0.66	0.99	0.89	0.9	0.0	0.0	0.0	0.0	0.0
650	1200	2500	0.54	0.96	0.98	0.95	0.88	0.00	0.88	0.99	0.70	0.0	0.9	0.0	0.0	0.0	0.0
300	2100	2600	0.97	0.96	0.63	0.00	0.59	0.98	0.66	0.99	0.89	0.0	0.0	0.9	0.0	0.0	0.0
400	1150	2500	0.99	0.99	0.84	0.99	0.84	0.00	0.88	0.99	0.70	0.0	0.0	0.0	0.9	0.0	0.0
500	2000	2750	0.90	0.99	0.96	0.00	0.72	0.99	0.22	0.95	0.99	0.0	0.0	0.0	0.0	0.9	0.0
400	900	2700	0.99	0.99	0.84	0.74	0.55	0.00	0.35	0.97	0.98	0.0	0.0	0.0	0.0	0.0	0.9
550	1800	2600	0.01	0.97	0.98	0.03	0.95	0.87	0.79	0.99	0.86	0.98	0.0	0.0	0.0	0.0	0.0
600	1400	2600	0	0.92	0.99	0.73	0.99	0.24	0.79	0.99	0.86	0.99	0.0	0.0	0.0	0.0	0.0

Table 2.6 Initial connection weights of FRGNN1 for Telugu vowel data in fuzzy case

Input to hidden layer (w_{lj})

0.0747	0.0832	0.0759	0.0849	0.0937	0.0892
0.0634	0.06202	0.0661	0.0713	0.0761	0.0757
0.0642	0.0655	0.0741	0.0868	0.0737	0.0798
0.1117	0.0948	0.1024	0.1079	0.1104	0.0862
0.0778	0.0893	0.0981	0.1114	0.1232	0.0962
0.1044	0.0890	0.1286	0.1106	0.1047	0.1266
0.0681	0.0768	0.0868	0.0899	0.0763	0.0925
0.0629	0.0620	0.0824	0.0697	0.0695	0.0845
0.0768	0.0713	0.0870	0.0941	0.0765	0.0855

Hidden to output layer (w_{kl})

0.0936	0.0648	0.0362	0.0166	0.1588	0.07930
0.0970	0.0615	0.0472	0.0082	0.1616	0.0675
0.1214	0.0881	0.0477	0.0237	0.1601	0.0904
0.1347	0.1101	0.0603	0.0172	0.1636	0.0780
0.1294	0.0813	0.0532	0.0197	0.1629	0.0836
0.1371	0.095	0.0463	0.0179	0.1663	0.0844

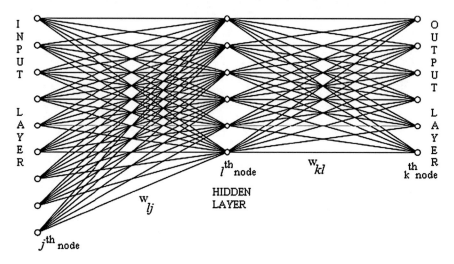

Fig. 2.7 The architecture of FRGNN1 for Telugu vowel data

The performance of the FRGNN1 is seen to vary over different folds. For example, in the 10-fold cross validation, for Telugu vowel data with initial weights in random case, crisp case and fuzzy case, the recognition scores of FRGNN1 are seen to vary between 79.01% (minimum) and 88.51% (maximum) with an average accuracy of 84.75%; 81.61% (minimum) and 90.80% (maximum) with an average accuracy of 86.55%; and 85.06% (minimum) and 94.25% (maximum) with an average accuracy of 87.71%, respectively. Similar type of recognition scores of FRGNN1 for remaining three data sets can also be found.

The generalization ability of the FRGNN1 is seen to depend on the characteristics of the data. If the correlation among input variables is high, the generalization ability is not so high and vice versa. For example, the generalization ability of the FRGNN1 for Telugu vowel, sonar and lymphography is below 90% because of the high correlation in the feature space of input vectors whereas it is above 90% for thyroid.

From Table 2.7, we can conclude that the performance of the FRGNN1 with initial connection weights corresponding to both fuzzy and crisp cases is superior to fuzzy MLP (i.e., FRGNN1 with initial weights corresponding to the random case) for all the data sets. Weight selection by Method II of Sect. 2.8 (fuzzy case) results in better performance than Method I of Sect. 2.8 (crisp case). Further, their difference is more apparent for overlapping classes of lymphography data (see Fig. 2.6a) where the difference is seen to be more than 3%.

Figure 2.8 presents the variation of the squared error with the number of epochs carried out during training of the FRGNN1 with initial weights in random, crisp, and fuzzy cases. This comparative result is shown, as an example, only for Telugu vowel data. Here, the results correspond to one fold over ten folds. As expected, the squared error decreases with the number of epochs. The error drops significantly

Table 2.7 Experimental results of FRGNN1

Dataset name	Initial weights case	Accuracy of fold										Average accuracy
		Fold 1	Fold 2	Fold 3	Fold 4	Fold 5	Fold 6	Fold 7	Fold 8	Fold 9	Fold 10	
Telugu vowel	Random	82.76	85.06	83.91	85.06	85.06	86.21	86.21	79.01	88.51	85.06	84.75
	Crisp	85.06	85.06	90.8	87.36	87.36	86.21	85.06	81.61	88.51	88.51	86.55
	Fuzzy	86.21	86.21	94.25	89.66	86.21	86.21	86.21	85.06	88.51	88.51	87.71
Sonar	Random	86.96	86.96	73.91	86.96	91.3	82.61	82.61	86.96	86.96	–	85.03
	Crisp	91.3	91.3	73.91	86.96	91.3	82.61	82.61	86.96	91.3	–	86.47
	Fuzzy	95.65	91.3	78.26	86.96	100	86.96	78.26	86.96	95.65	–	88.89
Lymphography	Random	87.5	81.25	75	75	81.25	81.25	75	87.5	75	–	79.86
	Crisp	87.5	87.5	75.00	75.00	87.5	81.25	81.25	93.75	81.25	–	83.33
	Fuzzy	100	87.5	75	81.25	93.75	81.25	81.25	100	81.25	–	86.81
Thyroid	Random	95.45	95.45	95.45	95.24	95.45	86.36	90.48	100.00	100.00	100.00	95.39
	Crisp	95.45	95.45	95.45	95.24	95.45	90.48	95.24	100.00	100.00	100.00	96.28
	Fuzzy	95.45	95.45	95.45	95.24	95.45	90.48	95.24	100.00	100.00	100.00	96.28

Fig. 2.8 Comparison of squared errors of the FRGNN1 with initial weights in *1* Random case, *2* Crisp case, and *3* Fuzzy case for Telugu vowel data for various epochs

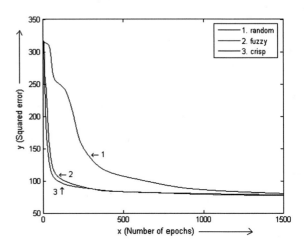

at lower number of epochs for the crisp and fuzzy cases than the random case of weight selection, because the former two cases enable the FRGNN1 to start learning from a much better position than the latter case. Among the crisp and fuzzy cases, although the crisp one is seen to provide slightly better results for the particular fold of vowel data (considered in Fig. 2.8), in overall performance, the fuzzy case of weight selection is the best for all the data sets.

Comparison of FRGNN1 with rough fuzzy MLP:

In order to compare the performance of FRGNN1 against an existing well known network of a rough fuzzy MLP (rfMLP) [1], we consider Telugu vowel data, as an example.

Table 2.8 depicts the performance of the rough fuzzy MLP, with one hidden layer as in the FRGNN1 with weights in crisp and fuzzy cases, for classification of vowels after 1500 epochs for a typical set of 10 folds. Parameters of the membership function (f_d and f_e) were assigned to the same values as in the FRGNN1 for the purpose of fair comparison. The values of learning rate (η), momentum parameter (α), and bias b are chosen by trail and error method. In rfMLP, we have obtained one reduct for each class representing its decision rule; thereby generating six decision rules for six vowel classes. By comparing the results of FRGNN1 with weights in both crisp case and fuzzy case, we can say that the performance of the FRGNN1 is better than rough fuzzy MLP.

In Fig. 2.9, we present the comparison of squared errors of the FRGNN1 (with initial weights in fuzzy case) and the rough fuzzy MLP for one of the 10-folds, considered in Table 2.8, for which both the FRGNN1 and rough fuzzy MLP have given the best classification accuracy within their respective sets of folds for vowel data. In the FRGNN1, the number of nodes in the hidden layer is equal to the number of classes which is six in case of vowel data. Interestingly, the same number of nodes was also obtained automatically in rough fuzzy MLP using its rough dependency factors. Again, the minimum values in both the cases were reached at 1500th epoch.

Table 2.8 Comparison of FRGNN1 and rough fuzzy MLP for Telugu vowel data

Dataset name	Initial weights case	Accuracy of fold										Average accuracy
		Fold 1	Fold 2	Fold 3	Fold 4	Fold 5	Fold 6	Fold 7	Fold 8	Fold 9	Fold 10	
Telugu vowel	FRGNN1 crisp case	85.06	85.06	90.8	87.36	87.36	86.21	85.06	81.61	88.51	88.51	86.55
	FRGNN1 fuzzy case	86.21	86.21	94.25	89.66	86.21	86.21	86.21	85.06	88.51	88.51	87.71
	rfMLP	81.61	83.91	87.36	88.51	87.36	87.36	87.36	83.91	88.51	89.66	86.55

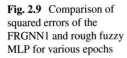

Fig. 2.9 Comparison of squared errors of the FRGNN1 and rough fuzzy MLP for various epochs

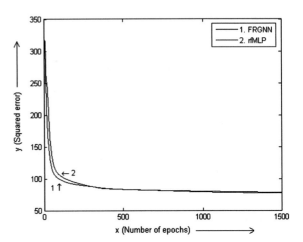

Considering the variation of average accuracy rate, Fig. 2.9 further supports the earlier findings concerning the superiority of FRGNN1 over rough fuzzy MLP for Telugu vowel data.

Salient Points of Difference between the FRGNN1 and rough fuzzy MLP:

In the rough fuzzy MLP, rough sets are integrated with a fuzzy MLP for encoding domain knowledge in network parameters whereas it is on fuzzy rough sets which are integrated with a fuzzy MLP in FRGNN1.

As stated in Method of rough fuzzy MLP, a threshold value (Tr) is used to eliminate the noisy patterns from input data. In contrast, no such threshold value is required in the FRGNN1 and the problem of removing noisy patterns is resolved by incorporating the concept of fuzzy rough sets in determining initial weights of the network.

In the rough fuzzy MLP, the decision rules are generated for each reduct of the reduct set and the dependency factors of these rules are mapped into the connection weights of the network. In contrast, no such attribute reducts are generated in the FRGNN1 and the dependency factors of all the conditional attributes and the average value of all the dependency factors of all conditional attributes, with respect to the decision classes, are defined in the form of initial connection weights of the network.

In the rough fuzzy MLP, network architecture is modeled for each reduct of the reduct set and every such network gives a recognition score for the test set. The maximum recognition score computed over them is then considered as the performance measure of the rough fuzzy for the concerned test set. In contrast, only one network architecture is modeled in the FRGNN1 corresponding to the entire training data, and the network is seen to perform better than the rough fuzzy MLP for the same test set.

Results of FRGNN2:

We perform 10-fold cross validation with stratified sampling. The FRGNN2 is trained using 9-folds of the data selected randomly from each of the classes and remaining one fold data is considered as the test set. This is repeated for 10 times

Table 2.9 Results of FRGNN2

Dataset name	Initial weights case	Accuracy of fold										Average accuracy
		Fold 1	Fold 2	Fold 3	Fold 4	Fold 5	Fold 6	Fold 7	Fold 8	Fold 9	Fold 10	
Image segmentation	Random	85.71	85.71	76.19	76.19	76.19	80.95	90.48	90.48	80.95	85.71	82.99
	Crisp	85.71	85.71	80.95	90.48	76.19	80.95	90.48	85.71	90.48	71.43	83.81
	Fuzzy	90.48	85.71	85.71	85.71	85.71	85.71	90.48	90.48	85.71	85.71	87.14
Sonar	Random	82.35	88.24	76.47	82.35	70.59	100	64.71	82.35	82.35	76.47	80.59
	Crisp	82.35	94.12	76.47	82.35	70.59	100	64.71	82.35	82.35	76.47	81.18
	Fuzzy	82.35	94.12	76.47	82.35	70.59	100	70.59	82.35	82.35	82.35	82.35
Spam base	Random	86.27	87.36	87.8	87.36	88.02	85.62	87.36	85.84	89.98	86.27	87.19
	Crisp	89.98	90.85	87.36	90.85	91.07	87.58	88.45	87.80	90.85	88.02	89.28
	Fuzzy	89.98	91.29	89.76	90.41	91.94	89.32	90.63	85.62	90.85	88.02	89.78

Table 2.10 Comparison of FRGNN2 and rough fuzzy MLP for Telugu vowel data

Dataset name	Initial weights case	Accuracy of fold										Average accuracy
		Fold 1	Fold 2	Fold 3	Fold 4	Fold 5	Fold 6	Fold 7	Fold 8	Fold 9	Fold 10	
Telugu vowel	FRGNN2											
	fuzzy case	88.24	85.88	89.41	88.24	83.53	84.71	85.88	85.88	88.24	87.06	86.71
	rfMLP	81.61	83.91	87.36	88.51	87.36	87.36	87.36	83.91	88.51	89.66	86.55

such that the entire data is employed in training and testing. The accuracies of 10-test sets and their average accuracy are provided in Table 2.9. The results of FRGNN2, as compared to FRGNN2 with weights in crisp case and fuzzy MLP [15] (FRGNN2 with weights chosen randomly within −0.5 to 0.5), are provided in Table 2.9.

The results from Table 2.9 show that FRGNN2 with weights in fuzzy case has achieved the best average classification accuracies for all data sets.

The comparative results of Telugu vowel data from (Table 2.10) reveal that the performance of FRGNN2 is superior to rfMLP. When the results of FRGNN2 are compared to FRGNN1 in fuzzy case (Table 2.8), the performance of FRGNN2 is inferior to FRGNN1. Therefore, it can be concluded that the FRGNN1 with weights in fuzzy case performs better than FRGNN2 and rfMLP when they are used for classification of patterns arising in the overlapping regions between the classes.

2.11 Conclusion

This chapter dealt with the problems of classification (supervised) in the framework of granular neural networks. Two models, namely FRGNN1 and FRGNN2, are described in detail. The concept of granulation is incorporated in both the input level and the initial connection weights of the network. The networks accept input in terms of fuzzy granules *low*, *medium* and *high*, and provides output decisions in terms of class membership values and zeros. They use the dependency factors of the conditional attributes, instead of random numeric connection weights, as the initial connection weights. Here, the dependency factors are determined using the concepts of fuzzy rough set. This investigation not only demonstrates a way of integrating fuzzy rough sets with a fuzzy neural network, but also provides a methodology that is capable of generating granular neural networks and improving their performance. The fuzzy rough sets provide an advantage that by which a discrete real-valued noise data can be effectively reduced without the need for any user supplied information (thresholds).

Two special types of granular computations are examined. One of them is induced by *low*, *medium* and *high* fuzzy granules and the other has classes of granulation structures induced by a set of fuzzy equivalence granules based on the fuzzy similarity relation/fuzzy reflexive relation. With respect to the classes of granulation structures, stratified fuzzy rough set approximations are obtained to determine the dependency factors of all conditional attributes to obtain the initial weights of the FRGNN1 and FRGNN2. The incorporation of granularity at various levels of the conventional MLP helps the resulting FRGNN1 and FRGNN2 to efficiently handle uncertain and ambiguous input information. This is demonstrated with extensive experimental results on a several real life data sets with varying dimension and size. The performance of FRGNN1 is found to be superior to rough fuzzy MLP which uses rough sets, rather than fuzzy rough sets, for knowledge encoding. Although, the comparative result is shown only for vowel data, the same observation holds for all other data sets considered in the experiment. The FRGNN1 and FRGNN2 models reflect the useful applications of granular computing to real world classification problems. Though their effectiveness has been demonstrated here on speech data, sonar data and medical data, these can be made well applicable to other real life problems such as gene expression analysis, web mining, social networks analysis, image and video analysis. Further, the information granules characterizing the fuzzy lower approximate regions of classes are used here for forming the basic networks. The effect of size and shape of the information granules on the network performances may therefore need to be investigated.

References

1. Banerjee, M., Mitra, S., Pal, S.K.: Rough fuzzy MLP: knowledge encoding and classification. IEEE Trans. Neural Netw. **9**(6), 1203–1216 (1998)
2. Cornelis, C., Jensen, R., Hurtado, G., Slezak, D.: Attribute selection with fuzzy decision reducts. Inf. Sci. **180**(2), 209–224 (2010)
3. Dick, S., Kandel, A.: Granular computing in neural networks. In: Pedrycz, W. (ed.) Granular Computing: An Emerging Paradigm, pp. 275–305. Physica Verlag, Heidelberg (2001)
4. Ganivada, A., Dutta, S., Pal, S.K.: Fuzzy rough granular neural networks, fuzzy granules, and classification. Theor. Comput. Sci. **412**, 5834–5853 (2011)
5. Ganivada, A., Pal, S.K.: A novel fuzzy rough granular neural network for classification. Int. J. Comput. Intell. Syst. **4**(5), 1042–1051 (2011)
6. Jang, J.S.R.: ANFIS: adaptive-network-based fuzzy inference systems. IEEE Trans. Syst. Man Cybern. **23**(3), 665–685 (1993)
7. Jensen, R., Shen, Q.: New approaches to fuzzy-rough feature selection. IEEE Trans. Fuzzy Syst. **17**(4), 824–838 (2009)
8. Klir, G.J., Folger, T.: Fuzzy Sets, Uncertainty and Information. Prentice Hall, N.J. (1988)
9. Mandal, D.P., Murthy, C.A., Pal, S.K.: Determining the shape of a pattern class from sampled points in R^2. Int. J. Gen. Syst. **20**, 307–339 (1992)
10. Mitra, S., De, R.K., Pal, S.K.: Knowledge-based fuzzy MLP for classification and rule generation. IEEE Trans. Neural Netw. **8**(6), 1338–1350 (1997)
11. Newman, D.J., Hettich, S., C. L. Blake, C.J.M.: UCI repository of machine learning databases, irvine. Department of Information and Computer Science, University of California, CA (1998). http://archive.ics.uci.edu/ml/
12. Pal, S.K., Majumder, D.D.: Fuzzy sets and decision making approaches in vowel and speaker recognition. IEEE Trans. Syst. Man Cybern. **7**(8), 625–629 (1977)
13. Pal, S.K., Majumder, D.D.: Fuzzy Mathematical Approach to Pattern Recognition. Wiley, New York (1986)
14. Pal, S.K., Mandal, D.P.: Linguistic recognition system based on approximate reasoning. Inf. Sci. **61**, 135–161 (1992)
15. Pal, S.K., Mitra, S.: Multilayer perceptron, fuzzy sets, and classification. IEEE Trans. Neural Netw. **3**(5), 683–697 (1992)
16. Radzikowska, A.M., Kerre, E.E.: A comparative study of fuzzy rough sets. Fuzzy Sets Syst. **126**(2), 137–155 (2002)

Chapter 3
Clustering Using Fuzzy Rough Granular Self-organizing Map

3.1 Introduction

Several clustering algorithms for identifying the actual structures of real life data sets are available in the literature. The clustering methods are used to determine groups where the patterns in the groups are similar and close to each other. Thus, the relationship among the groups of patterns can be analyzed. While the previous chapter deals with classification tasks and the initial connection weights of multilayer perceptron are defined in supervised manner, in this chapter self-organizing map (SOM) is used for clustering tasks where the same is determined in an unsupervised way. These are accomplished using fuzzy rough granular computing.

In general, the clustering methods do not use class information of data and hence they are called unsupervised classifiers. In this chapter, the self-organizing map, a neural network based clustering method, is used for all the data sets as it is a pattern discovery technique that not only clusters the data but also organizes the clusters for visualization. In SOM there are two layers, input and output, and nodes in the two layers are connected with initial weights. Different types of granular neural networks for clustering are formulated using the information granules and SOM. Here, the granules are developed using fuzzy sets, rough sets, fuzzy rough sets. The state-of-the-art SOM based granular neural networks include granular self-organizing map [6], fuzzy self-organizing map (FSOM) [11], lower and upper approximations based self organizing map [8] and rough reduct based self-organizing map [12]. The granular self-organizing map uses bidirectional propagation method for its training. The method [8], which organizes the web users collected from three educational web sites, is based on the concept of lower and upper approximations of a rough set. Fuzzy SOM [11] and rough reduct based self-organizing map [12] are based on hybridization of rough sets and fuzzy sets with SOM.

The aforesaid state-of-the-art methods are first explained in this chapter briefly. These are followed by a detailed description with experimental results of a recently developed method namely, fuzzy rough granular self organizing map (FRGSOM)

© Springer International Publishing AG 2017
S.K. Pal et al., *Granular Neural Networks, Pattern Recognition and Bioinformatics*, Studies in Computational Intelligence 712,
DOI 10.1007/978-3-319-57115-7_3

[5]. A fuzzy rough entropy measure developed using the concept of fuzzy rough sets and its properties are discussed.

The effectiveness of the FRGSOM and the utility of "fuzzy rough entropy" in evaluating cluster quality are demonstrated on different real life datasets, including microarrays, with varying dimensions. DNA microarray data allows analysis of large number of genes from different samples. It is made by a thin glass or nylon substrates or oligonucleotide arrays that contain thousands of spots representing genes. At first, mRNAs are extracted from tumor or normal cells (samples) and are converted into tumor cDNAs or normal cDNAs using reverse transcription. Transcription is a process wherein a genes forms ribonucleic acid (RNA) and RNA forms aminoacids through translation. The combination of tumor cDNA and normal cDNA can be made as complementary sequences of DNA. A scanner is used to measure the fluorescence values of tumor cDNA, normal cDNA and complementary sequences of DNA and those values are used as label values for different genes. The genes which contain the fluorescence values of complementary sequences of DNA can be treated as irrelevant genes or partially expressed genes. A group of genes that share similar expressions which is associated with a particular biological function is identified by clustering microarray data. Thus, functional relationships between groups of genes are identified.

3.2 The Conventional Self-organizing Map

In this section, first we provide a brief description of the conventional self organizing map (SOM) [7]. The self-organizing map contains two layers, namely, SOM's input layer and output layer (competitive layer). The number of nodes in the input layer is set equal to that of features of a pattern/object. For the output layer, it is set to be the number of clusters expected a prior. The weights to links connecting the nodes in the input and the hidden layers are initialized with a small set of random real numbers. The procedure for training the SOM is explained as follows:

Let, $x = x(e) \in R^n$ denote a sequence of input vectors and $\{w_{kj}(e), k = 1, 2, \ldots, c; j = 1, 2, \ldots, n\}$ denote a set of network parameters (initial connection weights) where e is number of iterations/time constant, c is the number of nodes in the output layer, and n represents the dimension of the input vector as well as the number of nodes in the input layer. Initially, the network parameters in SOM are chosen as small random numbers. At each successive instant of time, e, an input vector $x_j(e)$ is randomly presented to the network. The Euclidean distance, d_k, between the input vector, $x_j(e)$, and weight vector, $w_{kj}(e)$, is computed as

$$d_k = \|x_j(e) - w_{kj}(e)\|^2. \tag{3.1}$$

One of the output neurons which has the minimum Euclidean distance, is considered as a winning neuron, say winner, denoted by v and is expressed as

$$v = argmin\{d_k\}, k = 1, 2, \ldots, c, \tag{3.2}$$

The distance between a winning neuron (v) and an output neuron (k), represented by $\bigwedge_{vk}$, is given as

$$\bigwedge_{vk}(e) = \| r_v - r_k \|^2, \tag{3.3}$$

where r_v and r_k are the positions of a winning neuron (v) and an output neuron (k), respectively. By using $\bigwedge_{vk}$, the Gaussian neighborhood of a winning neuron, denoted by N_v, is defined as

$$N_v(e) = exp(\frac{-\bigwedge_{vk}^2}{2\sigma(e)^2}), \tag{3.4}$$

where $\sigma(e)$ represents the variance for controlling the width of the Gaussian neighborhood at iteration e. The value of σ is determined as

$$\sigma(e) = \sigma_0 exp(\frac{e}{\tau_1}), e = 0, 1, 2, \ldots, \tag{3.5}$$

where σ_0 is a initial value of σ and τ_1 is time constant. The value of σ will be decreased with different iterations and the weights of a winning neuron and the neighborhood neurons are updated as

$$w_{kj}(e + 1) = \begin{cases} w_{kj}(e) + \alpha(e)N_v(e)(x_j(e) - w_{kj}(e)), \\ \qquad\qquad if \ k \in N_v(e), \\ w_{kj}(e), else. \end{cases} \tag{3.6}$$

Here, α is a learning parameter, chosen between 0 and 1, as shown in [7] and the weights of the neurons that are not included in the neighborhood of the winning neuron are not modified. Equation 3.6 updates the weights of winning neuron and its neighborhood neurons. The updated weights are more likely to become similar to the input patterns which are presented to the network during training.

3.3 Granular Self-organizing Map

As described in [6], self-organizing map is formed by a set of granules. Each granule $g_i \in G_\phi$ can be decomposed into a new set of granules G_ψ, where $G_\phi \succeq G_\psi$. The set of granules G_ψ is called the child granule set of the parent granule $g_i \in G_\phi$. The construction method forms a granule g_i from a set of derived sub granules as

$$g_i = \frac{1}{|G_\psi|} \sum_{k=1}^{|G_\psi|} g_k, \forall g_k \preceq g_i, g_k \in G_\psi. \tag{3.7}$$

The granular self-organizing map [6] is described as a set of granules in a hierarchy $G_\phi = \{G_{\psi 1}, \ldots, G_{\psi t}\}$, where $G_{\psi 1} = \{G_{i,1}, \ldots, G_{i,m}\}$ is a set of self-organizing maps (SOMs) and t is the total depth of the hierarchy. Let $G_{\psi j} = \{g_1, \ldots, g_n\}$ be a SOM with n granules. For each $g_k \in G_{\psi j}$, it contains a storage unit s_k and weight vector w_k. Each granule has the structure $g_k = (w_k, s_k)$. The learning algorithm of granular self organizing map is described as follows:

Bidirectional update propagation: A SOM is trained on a set of input vectors. A similarity measure is applied between each granule and the input vectors to find a granule g_i, which has a weight vector closest to an input vector x_k. A granule g_i^* is marked as the winner for input vector x_k if it has the highest similarity value to the input vector. Once a winning granule has been identified, its weight vector must be updated according to the learning rate α. The value of α decays over time according to an iteration t. This ensures that the granules learn features quickly during training. This process is performed by using

$$g_i^*(t) = g_i^*(t-1) + \alpha(x_k(t) - g_i^*(t-1)), \tag{3.8}$$

where

$$g_i^* = argmin \sum_{j=1}^{m} (g_{ij} - x_{kj})^2. \tag{3.9}$$

Here, g_i^* signifies the degree of similarity between the input vector and a granule. The neighborhood set, $N_i^*(d)$, is calculated around g_i^* according to the decaying neighborhood distance d. A modified learning rate α' is used on the granules within the neighborhood set $N_i^*(d)$. The neighborhood granules of the winning neuron are then updated using

$$g_{N_i^*(d)}(t) = g_{N_i^*(d)}(t-1) + \alpha'(x_k(t) - g_{N_i^*(d)}(t-1)). \tag{3.10}$$

Further details of the bidirectional update propagation algorithm is available in [6].

3.4 Rough Lower and Upper Approximations Based Self Organizing Map

The architecture of this rough set based self organizing map [8] is similar to that of the Kohonen's self organizing map. Here, the map consists of two layers, an input layer and the Kohonen rough set layer (rough set output layer). These are fully connected. Each input layer neuron has a feed forward connection to each output layer neuron. Figure 3.1 illustrates the architecture of Kohonen rough set neural network (rough set based self organizing map). As discussed in [8], a neuron in the Kohonen layer

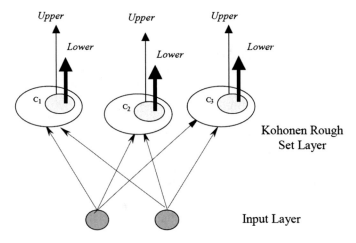

Fig. 3.1 Modified Kohonen neural network based on rough set theory [8]

consists of a lower neuron and an upper neuron. The lower neuron has an output of 1, if an object belongs to the lower bound of the cluster. Similarly, a membership in the upper bound of the cluster will result in an output of 1 from the upper neuron. When the lower and upper neurons have outputs of 1, then an object belongs to the lower bound of a cluster which also belongs to its upper bound. The description of an object belonging to the lower and upper bounds based on its membership values is elaborated in [8].

It is explained that the modification of the Kohonen's algorithm for obtaining rough sets to determine whether an object belongs to the upper or lower bounds of a cluster is based on its membership value. For every object, x, let $d(x, w_i)$ be the distance between itself and the weight vector w_i of cluster C_i. The ratios $\frac{d(x,w_i)}{d(x,w_j)}$, $1 \leq i, j \leq k$, are used to determine the membership of x as follows:

1. if $d(x, w_i)$ is the minimum for $1 \leq i \leq k$ and $\frac{d(x,w_i)}{d(x,w_j)} >$ threshold for any pair of weight vectors (w_i, w_j), then $x \in \overline{A}(w_i)$ and $x \in \overline{A}(w_j)$. Further, x is not part of any lower bound. The weight vectors w_i and w_j are modified as

$$w_i^{new} = w_i^{old} + \alpha_{upper}(t)(x - w_i^{old}),$$ (3.11)

$$w_j^{new} = w_j^{old} + \alpha_{upper}(t)(x - w_j^{old}).$$ (3.12)

2. When $x \in \underline{A}(w_i)$ and $d(x, w_i)$ is minimum for $1 \leq i \leq k$, the weight vector w_i is modified as

$$w_i^{new} = w_i^{old} + \alpha_{lower}(t)(x - w_i^{old}).$$ (3.13)

3.5   Fuzzy Self-organizing Map

In fuzzy self-organizing map (FSOM [11]), an n-dimensional input vector is transformed into a $3n$-dimensional linguistic input vector corresponding to the linguistic
terms *low*, *medium* and *high*. The nodes in the input layer of FSOM correspond to the
$3n$-dimensional linguistic input vector. In the output layer, the number of nodes is set
equal to the number of classes. The initial connection weights of FSOM are chosen
as random numbers within -0.5 to 0.5. The FSOM is trained using the competitive
learning algorithm.

3.6   Rough Reduct Based Self-organizing Map

The principle of the method [12] is follows: A set of fuzzy input vectors (training
patterns without considering their class labels) is first presented to an information
system, called attribute valued table. Its size is reduced using a heuristic threshold [1].
Based on the reduced attribute valued table, a discernibility function corresponding
to every pattern is defined. Attribute reducts are generated by applying conjunctive
and disjunctive normal operators on the discernibility functions. The reducts are then
used to determine initial connection weights of self-organizing map which is further
trained using competitive learning process.

The algorithm for determining the initial connection weights of rough self-
organizing map (Rough SOM) [12] is described as follows:

Step 1: An n dimensional input data is transformed into a $3n$-dimensional linguistic data, using π-membership functions, corresponding to linguistic terms *low*,
medium and *high*.

Step 2: Attributes which have membership values $\geq$ threshold value (0.5) are represented with 1 and others are represented with 0, in order to make a binary valued
data.

Step 3: The most representative pattern, i.e., a pattern which is repeated with maximum number of times, is selected from the resultant binary valued data to serve as
an object. Let there be m sets of objects, set_1, set_2, $\ldots$, set_m, in the attribute valued
table such that $n_{k_1} > n_{k_2} >, \ldots, n_{k_m}$ where $card(set_i) = n_{k_i}$ and $i = 1, 2, \ldots, m$.

Step 4: The m objects, selected from m different sets, are presented to the attribute
value table of size $m \times 3n$ where m represents the number of objects, and $3n$
represents the size of the $3n$-dimensional binary valued attributes.

Step 5: Let n'_{k_1}, n'_{k_2}, $\ldots$, n'_{k_m} denote the distant elements (cardinalities) among the
$n_{k_1}, n_{k_2}, \ldots, n_{k_m}$ such that $n'_{k_1} > n'_{k_2} > \ldots, > n'_{k_m}$.

Step 6: A heuristic threshold, Tr (using Eq. 2.13), is used to remove the objects from
the attribute valued table, which have the cardinalities less than the threshold value
Tr. The resulting attribute valued table is called a reduced attribute valued table,
say RS.

Step 7: Attribute reducts are generated from the reduced attribute valued table, RS. Let $U = \{x_1, \ldots, x_{m_1}\}$ and $\mathcal{A} = \{a_1, a_2, \ldots, a_{3n}\}$ be the set of objects and the set of binary valued attributes, respectively. A discernibility matrix, DM_{ij} of RS, of the size $m_1 \times m_1$, is computed as

$$DM_{ij} = \{a \in A : a(x_i) \neq a(x_j), \text{ for } i \text{ and } j = 1, 2, \ldots, m_1, \quad (3.14)$$

where $a(x_i)$ and $a(x_j)$ represent the attribute values of the patterns x_i and x_j, respectively. For each object $x_i \in U$, the discernibility function $f_{RS}^{x_i}(a_1, \ldots, a_{3n})$ is formulated as

$$f_{RS}^{x_i}(a_1, \ldots, a_{3n}) = \bigwedge \left\{ \bigvee (DM_{ij}) : 1 \leq j < i \leq m_1, DM_{ij} \neq \emptyset \right\}, \quad (3.15)$$

where $\bigvee (DM_{ij})$ is the disjunction of all attributes, called reducts.

In Rough SOM, the number of nodes in the input layer is determined corresponding to 3-dimensional features, and in the SOM output layer, it is set based on the number of reducts. Here, every node in the output layer is set corresponding to a reduct. The connection weights, between nodes of the input layer and the output layer of Rough SOM, are initialized with large random numbers for the attributes appearing in the reducts and with small random numbers for the attributes which did not appear in the reducts.

In the following sections, we describe in detail a fuzzy rough granular self-organizing map (FRGSOM [5]), recently developed, with comparative experimental results. This also includes a fuzzy rough measure and its properties.

3.7   Fuzzy Rough Granular Self-organizing Map

3.7.1   Strategy

In general, random numbers generated within 0 to 1 are used as initial connection weights of SOM [7]. The connection weights are not invariant for every new instant of training SOM. The SOM with random initial connection weights may result in slower convergence rate in the solution process. These weights may not be competent for a data with uncertain information. Moreover, the SOM uses 1-dimensional input vector (not granular input). To over come the situations, a fuzzy rough granular self-organizing map (FRGSOM [5]) is developed by determining 3-dimensional granular input vector and initial connection weights of SOM, using fuzzy sets and fuzzy rough sets, respectively. The problem of uncertainty, arising between the overlapping boundaries in a feature space, is handled using fuzzy sets and fuzzy rough sets. The concept of information granulation involving in the method of FRGSOM is described in three steps: (i) Each feature value is defined in terms of 3-dimensional granular input vector, *low*, *medium* or *high*, using the concepts of fuzzy set. (ii) Granulation structures (information granules) are based on a similarity matrix. It is computed by finding similarity among all possible pairs of patterns using fuzzy implication and t-norm. Then, the granulation structures are generated by applying an α-value (α-cut), between 0 and 1, on the similarity matrix. (iii) Membership values to the patterns belonging to lower approximation of a fuzzy set, representing an information granule, are computed. The FRGSOM is trained through the competitive learning process of the conventional SOM. After completion of the training process, the FRGSOM determines the underlying granulation structures/clusters of the data as its output.

3.7.2   Different Steps of FRGSOM

1. *Granulate the linguistic input data based on α-cut*:
   The linguistic input data is granulated in two phases. While the first phase computes the pairwise similarity matrix among the patterns using t-norm/T-norm and implication operator, the second phase generates the granulation structures. The value of α, used to generate granulation structures, is chosen between 0 to 1 in the second phase. The complete procedure of defining granular structures is explained in Sect. 3.7.3.

2. *Fuzzy rough sets to extract domain knowledge of data*:
   The granulation structures are first labeled according to decision class information and then presented to a decision table. Based on the decision table, a fuzzy reflexive relation is defined to measure a feature wise similarity between two patterns in the universe, thereby approximating each such structure using lower and upper approximations. The lower approximation is used to define a positive degree of a pattern in a universe and a dependency degree of a conditional attribute. The lower approximations, positive degrees, and dependency degrees are then used to extract the domain knowledge about the data. These are discussed in detail in Sect. 3.7.4.

3. *Incorporate the domain knowledge in SOM*:
   The domain knowledge about the data is extracted and is incorporated in the form of connection weights between the nodes of the input layer and the nodes of the output layer of self-organizing map (SOM).

4. *FRGSOM for clustering data*:
   The FRGSOM is trained with the competitive learning of SOM and the data is clustered. These are explained in Sect. 3.7.6.

5. *Example*:
   The knowledge encoding procedure and the network configuration of the FRGSOM are explained with an example in Sect. 3.7.5.

3.7.3 Granulation of Linguistic Input Data Based on α-Cut

The self-organizing map uses 3D input vector that is expressed in terms of linguistic terms *medium* and *high*. Each linguistic term can be treated as fuzzy granule. Here, the values of attributes/features of a pattern constitute an input vector. An n-dimensional vector is expressed as a $3n$-dimensional linguistic vector using Eq. 2.16 (see Sect. 2.7.1 and Chap. 2), based on π-membership function, Eq. 2.15 (see Sect. 2.20 and Chap. 2) where the centers and scaling factors of the membership function are chosen corresponding to fuzzy granules *low*, *medium* or *high* along each feature axis. The selection of the centers and scaling factors is explained in Sect. 2.7.2 (see Chap. 2).

The linguistic input data is granulated, to find the granulation structures of the data in two phases, using two different algorithms. While the first phase computes a pair wise similarity matrix of the size $s \times s$ where s is the total number of patterns (see Algorithm 1), the second phase generates the granulation structures (see Algorithm 2). Algorithm 1 is used to define a pair wise similarity matrix among the linguistic input data, using fuzzy logic connectives of Eqs. 2.24 and 2.26 (see Sect. 2.8, Chap. 2).

Algorithm 1. *Similarity Matrix*
1. **Input:** $x(i, j), i = 1, 2, \ldots, s; j = 1, 2, \ldots, 3n.$
2. /*a set of $3n$-dimensional granular data*/
3. **Output:** $m(i, i_1), i \& i_1 = 1, 2, \ldots, s.$
4. /*a similarity matrix*/
5. **Method:**

```
 1: for i ← 1 to s do
 2:   for i₁ ← 1 to s do
 3:     for j ← 1 to 3n do
 4:       X ← x(i, j);
 5:       Y ← x(i₁, j);
          /* Use Eq. 2.26 */
 6:       l₁ ← 1 − X + Y;
 7:       l₂ ← 1 − Y + X;
 8:       I₁ ← (l₁ < 1)?l₁ : 1;
 9:       I₂ ← (l₂ < 1)?l₂ : 1;
10:       /* Use Eq. 2.24 */
11:        M(j) ← ((I₁ + I₂ − 1) > 0)?(I₁ + I₂ − 1) : 0;
12:     end for j
13:     for j ← 1 to 3n do
14:        m(i, i₁) ← min[M(j)];
15:     end for j
16:   end for i₁
17: end for i
```

The similarity matrix is then used to develop the granulation structures based on an α-value where the value of α is chosen between 0 and 1. The method of determining granulation structures (i.e., p groups in Algorithm 2) is shown in the Algorithm 2.

Algorithm 2. *Granulation Structures*
1. **Input:** $x(i, j), i = 1, 2, \ldots, s;$
2. $j = 1, 2, \ldots, 3n.$ /*linguistic input data*/
3. $m(i, i_1), i, i_1 = 1, 2, \ldots, s.$
4. /*a similarity matrix*/
5. α /*a real value between 0 and 1*/
6. **Output:** p /* number of groups*/
7. p_{i_2} /*number of patterns in the group p*/
8. $U(i_2, i_3, j), i_2 = 1, 2, \ldots, p; i_3 = 1, 2, \ldots, p_{i_2}.$ /*granulation structures*/
9. **Method:**

    1: $p \leftarrow 0;$
    2: **for** $i \leftarrow 1$ to s **do**
    3:    /*use $< continue >$ statement*/
    4:    $p_{i_2} \leftarrow 0;$
    5:    $p \leftarrow p + 1;$
    6:    **for** $i_1 \leftarrow 1$ to s **do**
    7:      **if** $m(i, i_1) > \alpha$ **then**
    8:        /*use $< continue >$ statement*/
    9:        $flag(i_1) \leftarrow 1;$
    10:      $g(p, p_{i_2}) \leftarrow i_i;$
    11:      $p_{i_2} \leftarrow p_{i_2} + 1;$
    12:    **end if**
    13:    **end for**i_1
    14: **end for**i
    15: **for** $i_2 \leftarrow 1$ to p **do**
    16:    **for** $i_3 \leftarrow 1$ to p_{i_2} **do**
    17:      $val \leftarrow g(i_2, i_3);$
    18:      **for** $j \leftarrow 1$ to $3n$ **do**
    19:        $U(i_2, i_3, j) \leftarrow x(val, j);$
    20:      **end for** j
    21:    **end for**i_3
    22: **end for**i_2

The resultant structures (p groups) can be viewed as partitions or clusters. These partitions are arranged in decreasing order according to the size of the group. Here, the size is based on the number of points in a group. It may be noted that, for different α-values, between 0 and 1, the number of granulation structures will be different. We performed experiments with different α-values and for every α-value, we select the top c groups, based on their size, in all p groups where c represents the user defined number of clusters. The compactness of the first c groups, for every α-value, is then calculated using the fuzzy rough entropy (FRE) (defined in Sect. 3.8) and the granulation structures, corresponding to a particular α-value which provide the lowest average FRE are accepted. These are presented to a decision system S to extract the domain knowledge (explained in Sect. 3.7.4).

3.7.4 Fuzzy Rough Sets to Extract Domain Knowledge About Data

Let the number of patterns in all the c-groups, obtained using the selected α-value, be denoted by v. These v patterns from c-groups, represented by $\{x_1, x_2, \ldots, x_v\}$, are then presented to a decision system $SS = (U, \mathcal{A} \cup \{d\})$ where U represents the universe, $\mathcal{A}$ represents 3-dimensional conditional attributes, say $\{a_1, a_2, \ldots, a_{3n}\}$ and $\{d\}$ is a decision attribute. The decision classes, $d_k, k = 1, 2, \ldots, c$, are defined based on the c-groups, corresponding to the decision attribute. The set U is partitioned into a number of non-overlapping sets/concepts corresponding to the decision classes, say $d_k, k = 1, 2, \ldots, c$. The fuzzy reflexive relation R_a between any two patterns x and y in U with respect to a quantitative attribute $a \in \mathcal{A}$ is defined in Eq. 2.43 (see Sect. 2.9.1, Chap. 2).

For an attribute $a \in \{d\}$, the decision classes in fuzzy case are defined as

$$R_a(x, y) = \begin{cases} DD_{kk}, & \text{if } a(x) = a(y), \\ DD_{ku}, & \text{otherwise,} \end{cases} \tag{3.16}$$

for all x and y in U. Here DD_{kk} corresponds the membership values of all patterns in the kth class to its own class. It is defined as

$$DD_{kk} = \mu_k(\overrightarrow{F}_i), \text{ if } k = u, \tag{3.17}$$

where $\mu_k(\overrightarrow{F}_i)$ represents the membership value of the ith pattern to the kth class. The method for determination of $\mu_k(\overrightarrow{F}_i)$ is described in Sect. 2.8.1 (see Chap. 2).

The other, DD_{ku} corresponds the membership values of all patterns in the kth class to other classes and is defined as

$$DD_{ku} = 1, \text{ if } k \neq u \tag{3.18}$$

where k and $u = 1, 2, \ldots, c$.

The lower and upper approximations of a set $A \subseteq U$, based on fuzzy $\mathcal{B}$-indiscernibility relation $R_\mathcal{B}$ for all $a \in \mathcal{B} \subseteq \mathcal{A}$ and fuzzy decision classes R_d for all $a \in \{d\}$ of Eq. 3.16, are computed using Eqs. 2.34 and 2.35, respectively. The fuzzy positive region, based on the fuzzy $\mathcal{B}$-indiscernibility relation, is defined using Eq. 2.36. The dependency degree of a set of attributes $\mathcal{B} \subseteq \mathcal{A}, \gamma_\mathcal{B}$, is calculated using Eq. 2.38.

3.7.5   Incorporation of the Domain Knowledge in SOM

In this section, we first describe how the decision table can be used to explain the concept of granulation by partition and fuzzy rough set approximations, based on a fuzzy reflexive relation. Based on this principle, knowledge about the data is extracted and incorporated into the self organizing map (SOM). The resultant network is then trained using the competitive learning of SOM. The knowledge encoding procedure is as follows:

Knowledge encoding procedure: Based on the aforesaid decision table $SS = (U, \mathcal{A} \cup \{d\})$ with its set of conditional attributes, decision attributes, set of patterns, and labeled values of patterns corresponding to $3n$-dimensional conditional attributes, the following steps are applied to the aforesaid decision table $SS = (U, \mathcal{A} \cup \{d\})$ for extracting the domain knowledge about data.

(S1)   Generate a fuzzy reflexive relational matrix by using the fuzzy reflexive relation, using Eq. 2.43 (see Sect. 2.9.1), on all possible pairs of patterns and obtain additional granulation structures based on the relational matrix.

(S2)   Compute the membership value of a pattern belonging to lower approximation, using Eq. 2.34 (see Sect. 2.8.1), of a concept, based on the fuzzy reflexive relational matrix. Here, the fuzzy decision classes, Eq. 3.16, are involved in defining the lower approximations.

(S3)   Calculate the fuzzy positive region of a pattern in the concept, using Eq. 2.36 (see Sect. 2.8.1), corresponding to every conditional attribute.

(S4)   Compute the dependency degree of each conditional attribute, using Eq. 2.38 (see Sect. 2.8.1), based on the fuzzy positive regions of patterns within the concept, with respect to every decision class. Assign the resulting dependency factors as initial weights between the input layer nodes and c-number of output layer nodes of SOM where c represents the user defined number of clusters.

3.7.6   Training and Clustering

In the FRGSOM, the input data is first transformed into 3-dimensional granular space using Eq. 2.16 (see Sect. 2.7.1). During the training process of FRGSOM, instead of choosing the initial connection weights in SOM as small random numbers they are determined using the concept of fuzzy rough sets, explained in Sect. 3.7.5. The FRGSOM is then trained through the competitive learning process, explained in Sect. 3.2. After completion of the competitive learning, FRGSOM is able to partition the granular input data into the groups/clusters (granulation structures) in a topological order. Here, the input data is partitioned by FRGSOM in a topological order in the sense that the weight updates of the neighborhood neurons, of a winning neuron, cause the whole output space to become approximately ordered [7].

Table 3.1 Decision table with 3D conditional attributes and fuzzy decision classes

U	L_1	M_1	H_1	L_2	M_2	H_2	L_3	M_3	H_3	d	DD_{kk}	DD_{ku}
1	0.875	0.395	0	0.929	0.924	0	0.125	0.347	0	1	0.2517	1
3	0.5	0.697	0	0.382	0.739	0	0.125	0.875	0	1	0.2975	1
6	0	0.604	0.875	0	0.813	0.125	0	0.875	0.929	1	0.2198	1
2	0.875	0.395	0	0	0.260	0.125	0	0.986	0.382	2	0.2103	1
4	0	0.302	0.5	0.929	0.924	0	0	0.875	0.929	2	0.3211	1
5	0	0.604	0.875	0.929	0.92	0	0	0.875	0.929	2	0.3000	1

3.7.7 Examples

Let us consider the data set with two classes in Table 2.1 where the patterns in the two classes, say $A_1 = \{x_1, x_3, x_6\}$ and $A_2 = \{x_2, x_4, x_5\}$, denote two granulation structures. Every conditional attribute (feature) in the table is transformed into 3D granular space using Eq. 2.16 (see Sect. 2.7.1). Then the resulting decision table is shown in Table 3.1. The fuzzy decision classes, defined using Eq. 3.16, are also provided in the table under the decision attribute columns DD_{kk} and DD_{ku}.

Based on the procedure described in Sect. 3.7.5, we extract domain knowledge about data in Table 3.1 as follows. Using Step S1, the fuzzy reflexive relational matrix for every conditional attribute is computed. As an example, the relation matrices for the attributes L_1 and M_1 are provided in Sect. 2.9.2. (see Chap. 2). The membership value of pattern $x_1 \in A_1$, belonging to the lower approximation (using Eq. 2.34), with respect to the decision class d_1 is computed as

$$(R_{L_1} \downarrow R_d x)(x_1) = \inf_{x \in U} I\{R_{L_1}(x, x_1), R_{d_1}(x, x_1)\},$$
$$= \min\{I(1, 0.2517), I(0.938, 0.2975),$$
$$I(0.184, 0.2198), I(1, 1), I(0.194, 1),$$
$$I(0.194, 1)\},$$
$$= \min\{0.2517, 0.3594, 1, 1, 1, 1\},$$
$$= 0.2517.$$

For the remaining patterns, the membership values belonging to the lower approximation are $(R_{L_1} \downarrow R_d x)(x_3) = 0.2975$, $(R_{L_1} \downarrow R_d x)(x_6) = 0.2198$, $(R_{L_1} \downarrow R_d x)(x_2) = 0.2517$, $(R_{L_1} \downarrow R_d x)(x_4) = 0.2198$, and $(R_{L_1} \downarrow R_d x)(x_5) = 0.2198$.

The membership value of pattern x_1 belonging to the lower approximation (using Eq. 2.34) with respect to the decision class d_2 is

$$(R_{L_1} \downarrow R_d x)(x_1) = \inf_{x \in U} I\{R_{L_1}(x, x_1), R_{d_2}(x, x_1)\},$$
$$= \min\{I(1, 1), I(0.938, 1), I(0.184, 1),$$

$$I(1, 0.2103), I(0.194, 0.3211),$$
$$I(0.194, 0.3000)\},$$
$$= \min\{1, 1, 1, 0.2103, 1, 1\},$$
$$= 0.2103.$$

Similarly, for the remaining patterns, the membership values belonging to the lower approximation are $(R_{L_1} \downarrow R_d)(x_3) = 0.2715$, $(R_{L_1} \downarrow R_d x)(x_6) = 0.3000$, $(R_{L_1} \downarrow R_d x)(x_2) = 0.2103$, $(R_{L_1} \downarrow R_d x)(x_4) = 0.3000$, and $(R_{L_1} \downarrow R_d x)(x_5) = 0.3000$.

Hence, the positive regions (using Eq. 2.36) of patterns in the concepts $A_1 = \{x_1, x_3, x_6\}$ and $A_2 = \{x_2, x_4, x_5\}$ are defined as

$$POS_{L_1}(x_1) = \max\{0.2517, 0.2103\},$$
$$= 0.2517.$$

Similarly, for the remaining patterns in the concepts, we have

$$POS_{L_1}(x_3) = 0.2975, (R_{L_1} \downarrow R_d x)(x_6) = 0.3000.$$
$$POS_{L_1}(x_2) = 0.2517, POS_{L_1}(x_4) = 0.3000, \text{ and } POS_{L_1}(x_5) = 0.3000.$$

The dependency degree of the attribute L_1 with respect to a decision class d_1 is computed as

$$\gamma_{\{L_1\}}(x_0) = (0.2517 + 0.2975 + 0.3000)/3,$$
$$= 0.2831.$$

The dependency degrees for the remaining attributes with respect to each decision class are as follows:

$$\gamma_{\{M_1\}}(A_1) = 0.3142, \qquad \gamma_{\{M_1\}}(A_2) = 0.2981,$$
$$\gamma_{\{H_1\}}(A_1) = 0.2678, \qquad \gamma_{\{H_1\}}(A_2) = 0.2909,$$
$$\gamma_{\{L_2\}}(A_1) = 0.3037, \qquad \gamma_{\{L_2\}}(A_2) = 0.2732,$$
$$\gamma_{\{M_2\}}(A_1) = 0.3899, \qquad \gamma_{\{M_2\}}(A_2) = 0.4563,$$
$$\gamma_{\{H_2\}}(A_1) = 0.2732, \qquad \gamma_{\{H_2\}}(A_2) = 0.2732,$$
$$\gamma_{\{L_3\}}(A_1) = 0.7320, \qquad \gamma_{\{L_3\}}(A_2) = 0.2198,$$
$$\gamma_{\{M_3\}}(A_1) = 0.4754, \qquad \gamma_{\{M_3\}}(A_2) = 0.3081,$$
$$\gamma_{\{H_3\}}(A_1) = 0.7666, \text{ and } \qquad \gamma_{\{H_3\}}(A_2) = 0.3038.$$

Let us now define the initial structure of FRGSOM with the above said example dataset in Table 3.1. The data has nine input features (conditional attributes), so the number of input layer nodes of the FRGSOM is set to nine. The number of output layer nodes of the FRGSOM is set to two as it is assumed that there are two granulation structures in this data. The aforesaid nine dependency degrees, corresponding to the set A_1, are used as initial connection weights between the nine input layer nodes and the first node in the SOM's output layer. Similarly, the other nine dependency degrees, corresponding to the set A_2, are used as initial connection weights between the nine input layer nodes and the second output node. The FRGSOM is then trained through a competitive learning of SOM (explained in Sect. 3.2) for clustering the input data.

3.8 Fuzzy Rough Entropy Measure

The output clusters of FRGSOM are used for defining a entropy measure, called fuzzy rough entropy, based on the lower and upper approximations of a fuzzy set. In general, in real life data sets, pattern classes have overlapping boundaries, resulting in uncertainty with respect to class belongingness of patterns. The uncertainty can be handled by defining degrees of membership of the patterns, belonging to lower and upper approximations of a set, corresponding to each cluster. Several researchers have defined entropy measures, based on fuzzy set theory and rough set theory, in the past few years. Different classes of entropy measures using fuzzy rough sets are described in [10, 16]. The fuzzy rough entropy measure uses the lower and upper approximations of a fuzzy rough set. It quantifies the uncertainty in each cluster.

Before defining the fuzzy rough entropy measure, the concept behind it is explained. Let us consider, as an example, three clusters, say, CL_1, CL_2 and CL_3. Let, s_1, s_2 and s_3 denote the number of patterns belonging to CL_1, CL_2 and CL_3, respectively. It may be noted that, the data used for evaluation of the clusters, based on fuzzy rough entropy measure, is defined in terms of membership values using Eq. 2.15 (see Sect. 2.7, Chap. 2) where the parameters in Eq. 2.15 are considered, typically, corresponding to a linguistic term *medium*.

Let, A_q denote the qth set, with all the patterns, corresponding to the cluster CL_q, $q = 1, 2$ and 3. That is, for $q = 1$, $A_1 = CL_1$. The entropy measure for a cluster CL_q is described on the basis of the roughness value of the set A_q which is as follows:

Roughness of a set A_1 in a universe U: Let $S_q = (U_q, \mathcal{A} \cup \{d\})$ be a decision system corresponding to a cluster CL_q, $q = 1, 2$ and 3. For $q = 1$, $S_1 = (U_1, \mathcal{A} \cup \{d\})$ represents the decision system for the set A_1 where $\mathcal{A} = \{a_1, a_2, \ldots, a_n\}$ represents the conditional attributes, and d ($d \notin \mathcal{A}$) represents a decision attribute. Here, universe $U_1 = s_1 \cup s_2 \cup s_3$. The patterns, s_1, are labeled with a value "1", representing the decision class 1, and all other patterns, $s_2 \cup s_3$ are labeled with an integer value "2", representing the decision class 2. The methodology, say Procedure 1, of defining roughness of the set A_1, using the concepts of fuzzy rough set, is as follows:

Procedure 1:

(S1) For a quantitative attribute $a \in \mathcal{A}$, we calculate the fuzzy reflexive relation using Eq. 2.43 (see Sect. 2.9.1, Chap. 2).

(S2) For a qualitative decision attribute $a \in \{d\}$, we compute fuzzy decision classes for the patterns s_1 and $s_2 \cup s_3$ (see Sect. 2.9.1, Chap. 2).

(S3) Let the n-dimensional vectors O_{kj} and V_{kj}, $j = 1, 2, \ldots, n$, denote the mean and standard deviation of the data for the kth class of the decision system S_1. The weighted distance of a pattern $\overrightarrow{F_i}$ from the kth class is computed by Eq. 2.29 (see Sect. 2.8.1, Chap. 2) where $k = 1$ and 2 (decision class 1 and decision class 2). The membership values of the ith pattern to the kth class is computed by Eq. 2.20 (see Sect. 2.7.3, Chap. 2).

(S4) The values of the patterns corresponding to the decision classes are defined in terms of average membership values. Average membership values are defined in two ways, namely, (i) by computing the average of the membership values over all the patterns in the kth class to its own class $(k = 1)$, and assigning it to each pattern, $\overrightarrow{F}_i$, in the kth decision class $(k = 1)$, and (ii) by computing the average of the membership values over all the patterns in the kth class $(k = 1)$ to the other class $(k = 2)$, and assigning it to each pattern $\overrightarrow{F}_i$ in other decision class $(k = 2)$. So the average membership value of all the patterns in the kth class to its own class is defined as

$$E_{kk} = \frac{\sum_{i=1}^{m_k} \mu_k(\overrightarrow{F}_i)}{|m_k|}, \text{ if } k = u, \tag{3.19}$$

where u represents the total number of classes.

The average membership values of all the patterns in the kth class $(k = 1)$ to the other decision class (say, $k = 2$) are computed as

$$E_{ku} = \frac{\sum_{i=1}^{m_k} \mu_u(\overrightarrow{F}_i)}{|m_k|}, \text{ if } k \neq u, \tag{3.20}$$

where $|m_k|$ indicates the number of patterns in the kth class. For a qualitative attribute 'a' $\in \{d\}$, the fuzzy decision classes are calculated as

$$R_a(x, y) = \begin{cases} E_{kk}, & \text{if } a(x) = a(y), \\ E_{ku}, & \text{otherwise,} \end{cases} \tag{3.21}$$

for all x and y in U_1.

The decision table S_1 along with fuzzy decision classes for a set A_1 corresponding to a cluster CL_1 is shown in Table 3.2, as an example.

Let $x_0 = \{s_1\}$ and $x_1 = \{s_2 \cup s_3\}$ denote two subsets of the universe U_1. For each conditional attribute (feature) $a \in \mathcal{B} \subseteq \mathcal{A}$, we now compute the membership values of the patterns in the subset x_0, for belonging to the lower and the upper approximations of A_1, using Eqs. 2.34 and 2.35, respectively, (see Sect. 2.8.1, Chap. 2). Thereafter, for each conditional attribute $a \in \mathcal{B} \subseteq \mathcal{A}$, we calculate the sum of weighted membership values of all the patterns in a subset x_0 in two ways:

Table 3.2 An example decision table for a set A_1 corresponding to a cluster CL_1

Patterns (U_1)	Conditional attributes (A)	Decision attribute (d)	Fuzzy decision classes	
			(E_{kk})	(E_{ku})
s_1	$a_1, a_2, \ldots, a_n$	Class 1	E_{11}, for k = 1	E_{12}, for k = 1, u = 2
s_2, s_3		Class 2	E_{22}, for k = 2	E_{21}, for k = 2, u = 1

(i) by multiplying the membership value of a pattern y to a subset x_0, with its membership value for belonging to a lower approximation of A_1. For $x \in U_1$, it is denoted by LS as

$$LS = \sum_{y \in x_0} m(y)(R_B \downarrow R_d)(y), \tag{3.22}$$

where $m(y)$ represents the membership value of a pattern y to a subset x_0, and $(R \downarrow Ax)(y)$ represents its membership value for belonging to the lower approximation of A_1 corresponding to a conditional attribute $a \in A$.

(ii) by multiplying the membership value of a pattern y in a subset x_0, with the membership value for belonging to the upper approximation of a pattern y in a subset x_0. For $x \in U_1$, it is represented by US as

$$US = \sum_{y \in x_0} m(y)(R_B \uparrow R_d)(y). \tag{3.23}$$

For a conditional attribute $a \in A = \{a_1, a_2, \ldots, a_n\}$, the LS and US then become LS_{a_q} and US_{a_q} for $q = 1, 2, \ldots, n$. Therefore, the roughness of the set A_1 corresponding to the cluster CL_1 is calculated by

$$R(A_1) = 1 - \frac{\sum_{q=1}^{n} LS_{a_q}}{\sum_{q=1}^{n} US_{a_q}}. \tag{3.24}$$

Here, $R(A_1)$ quantifies the uncertainty in terms of roughness in the set A_1, corresponding to the cluster CL_1. The fuzzy rough entropy (FRE) of a cluster CL_1, based on the roughness measure defined in Eq. 3.24 of the set A_1, is defined as

$$FRE(CL_1) = -R(A_1)log_e(R(A_1)/e). \tag{3.25}$$

For the remaining clusters CL_q, q = 2 and 3, we apply the same procedure for defining the FRE. Therefore, the fuzzy rough entropy (FRE) of a cluster CL_q is defined as

$$FRE(CL_q) = -R(A_q)log_e(R(A_q)/e), \tag{3.26}$$

where $q = 1, 2, \ldots, c$, and the average fuzzy rough entropy is defined as

$$FRE = \frac{1}{c}\left(\sum_{q=1}^{c} FRE(CL_q)\right). \tag{3.27}$$

The fuzzy rough entropy measure, $FRE(CL_q)$, $q = 1, 2, \ldots, c$, satisfies the following properties:

1. Nonnegativity: $FRE(C_q) \geq 0$ iff $R(A_q) \geq 0$.
2. Continuity: For all the values of $R(A_q) \in [0, 1]$, $FRE(CL_q)$ is a continuous function of $R(A_q)$.
3. Sharpness: The value of FRE (CL_q) is zero when $(R_B \downarrow R_d)(y) = (R_B \uparrow R_d)$ (y), for all $y \in A_q$.
4. Maximality and Normality: When lower approximation of a set, A_q, is zero then the roughness values of the set is equal to 1. This implies that the entropy measure $FRE(CL_q)$ attains the maximum value of unity.
5. Resolution: For any $CL_q^* \leq CL_q \Rightarrow FRE(CL_q^*) \leq FRE(CL_q)$.
6. Monotonicity: $FRE(CL_q)$ is a monotonic increasing function of $R(A_q)$.

As $FRE(CL_q)$, $q = 1, 2, \ldots, c$ satisfies all the above mentioned properties, the average fuzzy rough entropy (FRE) also satisfies them. A lower value of FRE for a cluster indicates that the cluster is good (in terms of compactness).

The clusters obtained by using SOM, FSOM and FRGSOM are evaluated by the fuzzy rough entropy (FRE), β-index [13] and Davies-Bouldin index (DB-index) [3] for different values of c. A higher value of β-index indicates that the clustering solutions are compact whereas, it is opposite for FRE and DB-index.

3.9   Experimental Results

The clustering algorithms like SOM, FSOM, Rough SOM and FRGSOM are implemented in C language and all the programs are executed on Linux operating system installed in a HP computer. The computer is configured with Intel Core i7 CPU 880 @ 3.07 GHZ processor and 16 GB RAM. All the clustering algorithms are tested on different types of real-life data sets, namely, Telugu vowel [14] and gene expression microarray data sets like Cell Cycle [15], Yeast Complex [2, 4], and All Yeast [4].

The characteristics of gene expression data sets involving the names of microarray data sets, the number of genes, the number of time points (attributes), and the number of top level functional categories (classes) are provided in Table 3.3. Microarray data sets are often with missing gene expression values due to experimental problems. In this investigation, for Cell-Cycle data, out of 653 genes, 19 genes with missing gene expression values are first eliminated from the data set. Thereafter, the remaining 634 genes with all expression values are used in experiment. Similarly, for All Yeast data,

Table 3.3 Summary for different microarray data sets

Dataset	No. of genes	No. of time points	Classes
Cell Cycle	634	93	16
Yeast Complex	979	79	16
All Yeast	6072	80	18

out of 6221 genes, 6072 genes with all expression values are used in experiment. The Yeast Complex data has no missing gene expression values.

3.9.1 Results of FRGSOM

The output clustering solutions obtained using FRGSOM for every real life data set are described as follows.

Results of FRGSOM for Telugu vowel data: The Telugu vowel data is first transformed into a $3n$-dimensional ($n = 3$) granular space using Eq. 2.16 (see Sect. 2.7.1). If F_{i1}, F_{i2} and F_{i3} represent 1st, 2nd and 3rd features of ith pattern F_i, then the fuzzy granules of the features (in terms of *low*, *medium* and *high*) are quantified as

$$\overrightarrow{F}_{i1} \equiv \left\{ \mu_{low}(\overrightarrow{F}_{i1}), \mu_{medium}(\overrightarrow{F}_{i1}), \mu_{high}(\overrightarrow{F}_{i1}) \right\},$$

$$\overrightarrow{F}_{i2} \equiv \left\{ \mu_{low}(\overrightarrow{F}_{i2}), \mu_{medium}(\overrightarrow{F}_{i2}), \mu_{high}(\overrightarrow{F}_{i2}) \right\}, \text{ and}$$

$$\overrightarrow{F}_{i3} \equiv \left\{ \mu_{low}(\overrightarrow{F}_{i3}), \mu_{medium}(\overrightarrow{F}_{i3}), \mu_{high}(\overrightarrow{F}_{i3}) \right\}.$$

The transformed data is then used to find a similarity matrix using Algorithm 1. Algorithm 2 (see Sect. 3.7.3) generates the different numbers of groups for different values of α (0.2, 0.25, 0.35, 0.4, 0.45, 0.5, 0.55, 0.6, 0.65, 0.7, 0.75, 0.8) between 0 to 1. For example, 11 groups were found for $\alpha = 0.35$. These groups are then arranged in a descending order according to their size. The numbers of data points in these groups were 238, 211, 181, 94, 87, 31, 12, 8, 6, 2 and 1. Let c = 6 be the user defined number of clusters. We choose top 6 groups, based on their size, among all 11 groups. The FRE values for the 6 groups and the average FRE value are calculated. Similarly, we do the same process for the remaining values of α. Here, the average FRE of the first 6 groups is seen to be minimum for $\alpha = 0.35$. These 6 groups are the most compact and hence presented to a decision system SS. The knowledge about data in SS, is then extracted using the fuzzy rough sets and incorporated in FRGSOM.

During learning of the FRGSOM, 9 neurons are considered for the input layer as there are 3 input features in the vowel data and each feature has 3 dimensions. In the output layer of the SOM, we have considered 6 neurons corresponding to the 6 clusters. The initial connection weights, from the 9 nodes in the input layer

Table 3.4 Alignment of patterns at winning nodes in the output layer of FRGSOM for Telugu vowel data

$k1^{(100)}$	$2^{(171)}$	$3^{(191)}$
$4^{(62)}$	$5^{(143)}$	$6^{(204)}$

Table 3.5 Confusion matrix corresponding to data points at the winning nodes of the FRGSOM for Telugu vowel data

	δ	a	i	u	e	o
Cluster 1	43	11	2	3	41	0
Cluster 2	16	66	0	27	9	53
Cluster 3	0	0	145	0	46	0
Cluster 4	0	0	0	45	0	17
Cluster 5	10	0	25	0	106	2
Cluster 6	3	12	0	76	5	108

Table 3.6 Alignment of patterns at winning nodes in the output layer of SOM with random weights within −0.5 to 0.5 for Telugu vowel data

$1^{(104)}$	$2^{(178)}$	$3^{(154)}$
$4^{(233)}$	$5^{(156)}$	$6^{(46)}$

to the 6 nodes in SOM's output layer, are determined by dependency factors. The resultant network is then trained with competitive learning (see Sect. 3.2). After the completion of the learning process, the number of samples, obtained at each of the 6 output nodes of SOM, is shown in Table 3.4. For example, 100 data points are obtained at node 1 and 204 data points are obtained at node 6, in the SOM's output layer. The description of these clusters in a confusion matrix is shown in Table 3.5. From the results in Table 3.5, obtained with FRGSOM using fuzzy rough sets, the sum of the diagonal elements are found to be 513. The data points belonging to different clusters and the sum of the diagonal elements are seen to corroborate well with the actual class description (see Fig. 2.5a, Sect. 2.10).

Comparison of FRGSOM with SOM and FSOM: Tables 3.6 and 3.7 show the performance of SOM. The results of FSOM are shown in Tables 3.8 and 3.9.

It can be seen from Tables 3.7 and 3.9 that the sum of diagonal elements for SOM and FSOM are 441 and 502, respectively which are lower than that of FRGSOM. Let us now analyze the cluster wise results of FRGSOM, as compared to SOM and FSOM.

Cluster 1: It can be observed from Fig. 2.5a. that, the cluster 1, corresponding to class δ, has overlapping boundaries with the classes a, e and o. This information is also preserved and reflected in the results for cluster 1 using FRGSOM where out of 100

Table 3.7 Confusion matrix corresponding to data points at the winning nodes of the SOM random weights within −0.5 to 0.5 for Telugu vowel data

	δ	a	i	u	e	o
Cluster 1	41	14	2	1	45	1
Cluster 2	18	62	0	26	12	60
Cluster 3	0	0	110	0	44	0
Cluster 4	1	6	0	121	0	105
Cluster 5	4	0	57	0	94	1
Cluster 6	8	7	3	3	12	13

Table 3.8 Alignment of patterns at winning nodes in the SOM's output layer of FSOM with random weights within −0.5 to 0.5 for Telugu vowel data

$1^{(106)}$	$2^{(176)}$	$3^{(198)}$
$4^{(63)}$	$5^{(137)}$	$6^{(191)}$

Table 3.9 Confusion matrix corresponding to data points at the winning nodes of the FSOM with random weights within −0.5 to 0.5 for Telugu vowel data

	δ	a	i	u	e	o
Cluster 1	46	11	2	7	40	0
Cluster 2	13	66	0	25	9	63
Cluster 3	0	0	146	0	52	0
Cluster 4	0	0	0	45	0	18
Cluster 5	11	0	24	0	101	1
Cluster 6	2	12	0	74	5	98

data points (see node 1 in Table 3.4), 43, 11 and 41 data points belong to the classes δ, a and e (see Table 3.5), respectively. In contrast, using SOM, out of 104 data points at node 1 (see Table 3.6), 41, 14, 45 and 1 data points belong to the classes δ, a, e and o, respectively, i.e., only 41 points belong to class δ. Using FSOM, out of 106 data points at node 1 (see Table 3.8), 46, 11 and 40 data points belong to the classes δ, a and e, respectively, (see cluster 1 in Table 3.9), i.e., only 46 samples correspond to class δ. Although, the overlapping class information is reflected in the results for all the methods, the number of data points (100, 104 and 106, respectively) allotted in cluster 1 for all the methods are higher than the actual number of points (72) and FRGSOM is moderately better than related methods. The results for SOM and FSOM show that class δ does not have overlapping information with class o which is in contrast with the actual class information. Moreover, in SOM, the actual number of points (diagonal entry) belonging to class δ is less than those points belonging to class e whereas for a confusion matrix, the diagonal entry should be the dominant entry in each row.

Table 3.10 Comparison of β-index, DB-index and FRE values of clustering methods for Telugu vowel data

Algorithm	No. of clus. (c = 4)			No. of clus. (c = 6)			No. of clus. (c = 8)		
	β-index	DB index	FRE	β-index	DB index	FRE	β-index	DB index	FRE
FRGSOM	1.3045	1.1366	0.1688	1.1541	1.4687	0.1651	1.1116	1.6006	0.2163
FSOM	1.2032	1.1670	0.3342	1.0753	1.5256	0.3228	0.9827	1.7704	0.4220
SOM	0.9488	3.5593	0.5681	0.9317	1.5330	0.4452	0.9206	1.8450	0.4508

Cluster 6: Now consider cluster 6 where class o has overlapping boundaries with the classes δ, a, u and e (see Fig. 2). This information is supported by FRGSOM (see Table 3.5) where out of 204 data points at node 6 (see Table 3.4), 3, 12, 76, 5, and 108 data points belong to the classes δ, a, u, e, and o, respectively. In contrast, while using SOM, 46 data points are obtained at node 6 (see Table 3.6), and 8, 7, 3, 12, and 13 data points belong to the classes δ, a, u, e, and o, respectively (see Table 3.6). From the results of FSOM, we can observe that, 191 data points are obtained at node 6 (see Table 3.8), and 2, 12, 74, 5, and 98 data points belong to the classes δ, a, u, e, and o. Here, FRGSOM performs better than SOM and FSOM in terms of total number of data points and actual data points. Moreover, this shows the effectiveness of incorporating fuzzy rough sets in extracting domain knowledge. The results of SOM also show that the class o has overlapping boundary with the class i which is not correct. Similar conclusions on the FRGSOM, SOM and FSOM can also be made, based on results for the remaining clusters.

Evaluation of clusters: For a given value of c and a clustering method, all indices are computed on the resulting clustering solution and the results are reported in Table 3.10.

From Table 3.10, for $c = 4$, 6 and 8, we find that, the values of β-index and DB-index for FRGSOM are higher and lower, respectively, than those of FSOM and SOM. The FRE values for FRGSOM are also lower than those of FSOM and SOM. This signifies that, the FRGSOM performs the best in terms of all the indices. We can also observe that, for FRGSOM, FSOM and SOM, β-index and DB-index for $c = 4$ are higher and lower, respectively, than those values for $c = 6$ and 8. This means, $c = 4$ is the best choice for Telugu vowel data according to β-index and DB-index. In contrast, for all the clustering methods, FRE values are seen to be the lowest for $c = 6$. Note that, Telugu vowel data has 6 overlapping classes (shown in Fig. 2.5a.) and this information is therefore truly reflected by FRE.

As a typical example, now we compare the values of FRE for clusters obtained by SOM, FSOM and FRGSOM for $c = 6$. The results are shown in Table 3.11. The FRE values for cluster 1, obtained by SOM and FSOM, are 0.3950 and 0.6340, respectively. In contrast, the same for cluster 1, obtained by FRGSOM, is 0.1818, and it indicates that cluster 1 for FRGSOM is more compact than SOM and FSOM. Similar results are observed for remaining clusters, except cluster 5 where SOM

Table 3.11 Comparison of FRE values of SOM, FSOM and FRGSOM for $c = 6$

Clusters	SOM	FSOM	FRGSOM
1	0.3950	0.6340	0.1818
2	0.5479	0.3563	0.3300
3	0.2250	0.0350	0.0308
4	0.7609	0.1847	0.0858
5	0.2754	0.5478	0.2922
6	0.4673	0.1790	0.0981
Avg. FRE	0.4452	0.3228	0.1698

Table 3.12 Comparison of β-index, DB-index and FRE values of clustering methods for gene expression data sets

Algorithm	Cell Cycle			Yeast Complex			All Yeast		
	No. of clus. ($c = 16$)			No. of clus. ($c = 16$)			No. of clus. ($c = 18$)		
	β-index	DB index	FRE	β-index	DB index	FRE	β-index	DB index	FRE
FRGSOM	0.0663	13.2738	0.1278	0.0643	20.9827	0.1319	0.0561	41.7863	0.0364
FSOM	0.0631	14.6150	0.1864	0.0622	21.8147	0.1461	0.0548	42.3979	0.0564
SOM	0.0641	16.8528	0.2097	0.0648	24.7178	0.1525	0.0556	56.1796	0.0604

performs the best and FRGSOM is the second best. As per the average FRE, the FRGSOM is therefore seen to be the best.

Results of microarray data: The microarray gene expression data sets, like Cell Cycle, Yeast Complex and All Yeast are partitioned into different clusters using FRGSOM. The results are compared with SOM and FSOM in terms of β-index, DB-index and FRE. The values of indices are depicted in Table 3.12.

It can be seen from Table 3.12 that the values of DB-index and FRE are the smallest for FRGSOM as compared to FSOM and SOM for all the gene expression microarray datasets. The results signify that the performance of FRGSOM is better than the other two related clustering methods for all the gene expression datasets in terms of DB-index, FRE and β-index, expect for Yeast Complex data where only β-index for SOM is the best.

3.10 Biological Significance

After clustering the gene expression data using the FRGSOM, one or several functional categories are assigned to each cluster by calculating the P-values for different functional categories in Munich Information for Protein Sequences (MIPS) [9]. Using hypergeometric distribution, the probability (P-value) of observing at least m genes from a functional category within a cluster of size n is given by

$$P = 1 - \sum_{i=0}^{m-1} \frac{\binom{f}{i}\binom{N-f}{n-i}}{\binom{N}{n}} \tag{3.28}$$

where f is the total number of genes within a functional category and N is the total number of genes within the genome. A lower P value for a functional category within a cluster indicates that the functionally related genes are grouped in a better way by the clustering method.

Experiments are performed for $c = 16$, 16 and 18 for Cell Cycle, Yeast Complex and All Yeast data sets, respectively. The gene expression datasets are initially transformed into a 3-dimensional granular space. During the learning of FRGSOM, the number of nodes in the SOM's input layer is set to 279, 237, and 240 and in the SOM's output layer, it is set to 16, 16 and 18 for Cell Cycle, Yeast Complex and All Yeast data sets, respectively. After completion of competitive learning of FRGSOM, microarray datasets are partitioned corresponding to the number of user defined cluster numbers. As an example, the results of top 10 number of clusters, out of 16 clusters, for Cell Cycle data are presented in Table 3.13. Here, the number of genes in each cluster, the name of the functional category (obtained from the most significant P-value using Eq. 3.28), the actual number of genes within the category, the actual number of genes within the genome, and the functional category related P-values are shown in the table. The results for other clusters and datasets are available in [5].

Using the clustering solutions of Cell Cycle data, obtained by FRGSOM, we found that, every cluster shows functional enrichment in cell-cycle, mitotic cell cycle, cell cycle control and mitotic cell cycle as the data is itself related with cell cycle. The lowest and highest P-values involved in these functional categories are $2.38e^{-76}$ and $1.33e^{-03}$, respectively. The functional enrichment in other categories, other than Cell Cycle, is also shown in Table 3.13. From Table 3.13, we can observe that 10 clusters have more than 28 genes and show functional enrichment in more than one category. As a typical example, cluster 1 have 102 genes and shows functional enrichment in phosphate metabolism, modification by phosphorylation dephosphorylation, autophosphorylation, cytoskeleton/structural proteins, and fungal and other eukaryotic cell type differentiation category with p-values $8.05e^{-06}$, $2.29e^{-09}$, $2.20e^{-10}$ and $7.67e^{-09}$, respectively.

The biological significance of the FRGSOM is also compared with SOM and FSOM, using gene expression microarray data sets, like Cell Cycle, Yeast Complex and All Yeast. The functional enrichment results for SOM and FSOM are not provided here but, we present a comparison among different clustering methods in terms of NP value, obtained from functional enrichment of clustering solutions. The NP is defined as

$$NP = \prod_{i=1}^{n} (1 - P_i), \tag{3.29}$$

Table 3.13 Results for Cell Cycle microarray data using FRGSOM

Cluster no.	No. of genes within cluster	MIPS functional category	Category related genes within cluster	No. of within genome	p-value
1	102	Phosphate metabolism	21	418	$8.05e^{-06}$
		Modification by phosphorylation, dephosphorylation, autophosphorylation	18	186	$2.29e^{-09}$
		Cytoskeleton/structural proteins	22	252	$2.20e^{-10}$
		Fungal and other eukaryotic cell type differentiation	27	452	$7.67e^{-09}$
2	87	Modification by phosphorylation, dephosphorylation, autophosphorylation	8	186	$9.62e^{-04}$
		Microtubule cytoskeleton	9	47	$1.44e^{-09}$
		Fungal and other eukaryotic cell type differentiation	16	452	$2.39e^{-05}$
3	65	Phosphate metabolism	12	418	$2.57e^{-04}$
		Cell fate	9	273	$6.32e^{-04}$
		Fungal and other eukaryotic cell type differentiation	13	452	$1.32e^{-04}$
4	49	Modification by phosphorylation, dephosphorylation, autophosphorylation	8	186	$9.59e^{-05}$
		Cell fate	11	273	$7.19e^{-06}$
		Cytoskeleton/structural proteins	11	252	$3.31e^{-06}$
		Budding, cell polarity and filament formation	13	313	$6.21e^{-07}$
5	47	DNA damage response	5	77	$3.08e^{-04}$
		Cytoskeleton/structural proteins	8	252	$6.60e^{-04}$
		Development of asco- basidio- or zygospore	9	167	$4.49e^{-06}$
6	36	Kinase inhibitior	2	14	$4.34e^{-03}$
		Small GTPase mediated signal transduction	5	58	$5.21e^{-05}$
		Budding, cell polarity and filament formation	8	313	$1.49e^{-03}$
7	33	Protein binding	8	392	$1.33e^{-03}$
		Microtubule cytoskeleton	4	47	$1.33e^{-04}$
8	33	Phosphate metabolism	9	418	$2.58e^{-04}$
		Modification by phosphorylation, dephosphorylation, autophosphorylation	8	186	$4.48e^{-06}$
		Fungal and other eukaryotic cell type differentiation	12	452	$1.90e^{-06}$

Table 3.13 (continued)

Cluster no.	No. of genes within cluster	MIPS functional category	Category related genes within cluster	No. of within genome	p-value
9	30	Kinase inhibitior	2	14	$2.17e^{-03}$
		Pheromone response, mating-type determination	7	189	$3.33e^{-05}$
		Fungal and other eukaryotic cell type differentiation	11	452	$6.78e^{-06}$
10	29	Mating (fertilization)	3	69	$4.46e^{-03}$
		Cytoskeleton/structural proteins	9	252	$1.95e^{-06}$
		budding, cell polarity and filament formation	7	313	$6.22e^{-04}$

Table 3.14 Comparison of FRGSOM with Rough SOM, in terms of β-index, DB index, fuzzy rough entropy (FRE) and f_c, for Telugu vowel. The results are shown for 6 clusters as Rough SOM provides 6 clusters only

Data set	Algorithm	Iterations	β-index	DB index	FRE	f_c	CPU time in sec.
Telugu vowel data	FRGSOM	80	1.1541	1.4687	0.1690	513	3.0740
	Rough SOM	132	0.9911	1.6644	0.1699	464	2.1177

where P_i represents the most significant P-value (lowest one) associated with the ith cluster, and n represents the total number of clusters obtained using any clustering method. Here, $(1 - P_i)$ gives the probability that *not* observing the related functional enrichment in ith cluster and NP represents the probability that *not* observing any functional enrichment in all the clusters, found by any clustering method. Using Cell Cycle data, the values of NP for FRGSOM, SOM and FSOM are found to be 0.99, 0.94, and 0.95, respectively. The NP values for Yeast Complex and All Yeast microarray data sets are found to be 0.98 and 0.99 for FRGSOM, 0.96 and 0.97 for FSOM, and 0.96 and 0.95 for SOM. For all of the gene expression data sets, the NP values for FRGSOM are found greater than the related methods and indicate that, FRGSOM partitions the functionally related genes in a better fashion.

Comparison of FRGSOM with Rough SOM: Table 3.14 shows the performance of FRGSOM and Rough SOM in terms of β-index, DB-index and fuzzy rough entropy (FRE). For both the methods, the results are shown on Telugu vowel data for 6 clusters. The value corresponding to f_c in Table 3.14 represents the sum of diagonal elements in the confusion matrix for Telugu vowel data.

Table 3.15 Comparison of β-index, DB-index and FRE values of clustering methods for Cell Cycle data

Algorithm	Cell Cycle			CPU time in sec.
	No. of clusters $c = 6$			
	β-index	DB index	FRE	
FRGSOM	0.1988	8.5903	0.1270	9.9282
RSOM	0.1845	10.5134	0.1302	8.4490
FSOM	0.1711	18.6188	0.1472	9.0630
SOM	0.1660	18.0185	0.1983	7.7760

From the results in Table 3.14, we observe that FRGSOM performs better than Rough SOM for Telugu vowel data in terms of β-index and DB-index whereas FRE values of FRGSOM and Rough SOM are seen to be almost same. The f_c values for Telugu vowel data indicate that, the total number of points correctly classified by FRGSOM is much higher than that of RSOM (e.g., 513 vs. 464). It can also be observed that the CPU time for FRGSOM is higher than Rough SOM for the data.

We obtain 6, 12 and 14 reducts for Cell-Cycle, Yeast Complex and All Yeast data sets, respectively, by using Rough SOM where each reduct corresponds to one cluster. Hence, for the purpose of comparison, the performance of FRGSOM, FSOM and SOM are also tested on same numbers of clusters (6, 12 and 14) for those data sets. The results of β-index, DB-index and fuzzy rough entropy (FRE) for all the methods and a typical Cell Cycle data are shown in Table 3.15. From Table 3.15, it can be observed that β-index for FRGSOM is higher than that of RSOM, FSOM, SOM, and DB index & FRE values for FRGSOM are less than that of RSOM, FSOM and SOM. This implies that FRGSOM outperforms than the other methods. Similar type of results are also obtained for Yeast Complex and All Yeast data sets. The CPU time for FRGSOM and all the data sets is found to be higher than the remaining methods (see Table 3.15 for a typical Cell Cycle data). Since, the FRGSOM uses the initial connection weights obtained using fuzzy rough sets, instead of random numbers within -0.5 to 0.5.

Salient points in difference between the FRGSOM and the Rough SOM

1. In Rough SOM, the network is mainly modeled by the integration of the fuzzy sets and rough sets with SOM. In contrast, the FRGSOM is developed by the integration of the fuzzy set theory and fuzzy rough sets with SOM.
2. In Rough SOM, rough set theory is used to extract attribute reducts from data and further each reduct is used to generate a set of rules. In contrast, fuzzy rough set theory is used in FRGSOM to generate rules in terms of dependency factors.
3. The number of nodes, in the output layer of Rough SOM, is based on the attribute reducts whereas the number of nodes in the output layer of the FRGSOM is set according to the class information of the data.
4. The initial connection weights, between nodes of the input and output layers of Rough SOM, are defined with large random numbers for attributes which

appeared in the rules, and for the attributes which did not appear in the rules, are defined with small random numbers. In contrast, no such attribute reducts are generated using fuzzy rough sets for defining the initial connection weights in the FRGSOM. Attribute dependency factors, determined using the concept of fuzzy rough sets, are used in FRGSOM as initial connection weights.

3.11 Conclusion

The problem of clustering (unsupervised classification) and the development of self organizing map (SOM) is dealt within this chapter. First of all, various state-of-the-art versions of SOM, including classical SOM, rough SOM and fuzzy SOM, are described. Then, the relevance of a fuzzy rough granular neural network (FRGSOM) and its detailed methodologies are explained. This involves an integration of the concepts of fuzzy rough set with SOM, to predict the underlying clusters in a dataset and to handle the uncertainty that comes from patterns of the overlapping regions. Fuzzy rough sets, based on a fuzzy reflexive relation, are used to extract the domain knowledge in the form of dependency factors of conditional attributes. The dependency factors are encoded as the initial weights of the network, and the input vector of the network is defined in terms of fuzzy granules *low*, *medium* and *high*. The superiority of the FRGSOM, as compared to SOM, FSOM, and Rough SOM, is extensively demonstrated on Telugu vowel and microarray gene expression data sets, with dimension varying from 3 to 93.

A new fuzzy rough entropy measure (FRE) is also described using the concept of fuzzy rough sets, based on a fuzzy reflexive relation. This measure is capable of quantifying the uncertainties arising from overlapping regions. The lower value of entropy of a cluster signifies that the compactness of the cluster is high. The FRE is seen to reflect well the actual number of clusters for a dataset as compared to β-value and DB-index. FRGSOM is not only seen to be superior, but it also demonstrates a way of integrating two different facets of natural computing, namely, granular computing, self-organization in soft computing framework.

Here, fuzzy information granules are used in the processing stage to find homogeneous clusters. The concept of granules can be further used in decision level in different degrees of granularity in finding sub clusters. In this context, the effect of their sizes and shapes may also be investigated.

References

1. Banerjee, M., Mitra, S., Pal, S.K.: Rough fuzzy MLP: knowledge encoding and classification. IEEE Trans. Neural Netw. **9**(6), 1203–1216 (1998)
2. Bar-Joseph, Z., Gifford, D.K., Jaakkola, T.S.: Fast optimal leaf ordering for hierarchical clustering. Bioinformatics **17**(1), 22–29 (2001)

3. Davies, D.L., Bouldin, D.: A cluster separation measure. IEEE Trans. Pattern Anal. Mach. Intell. **1**(2), 224–227 (1979)
4. Eisen, M.B., Spellman, P.T., Brown, P.O., Botstein, D.: Cluster analysis and display of genome-wide expression patterns. Proc. Natl. Acad. Sci. USA **95**, 4863–14868 (1998)
5. Ganivada, A., Ray, S.S., Pal, S.K.: Fuzzy rough granular self-organizing map and fuzzy rough entropy. Theoret. Comput. Sci. **466**, 37–63 (2012)
6. Herbert, J.P., Yao, J.T.: A granular computing frame work for self-organizing maps. Neurocomputing **72**(13–15), 2865–2872 (2009)
7. Kohonen, T.: Self-organizing maps. Proc. IEEE **78**(9), 1464–1480 (1990)
8. Lingras, P., Hogo, M., Snorek, M.: Interval set clustering of web users using modified kohonen self-organizing maps based on the properties of rough sets. Web Intell. Agent Syst. **2**(3), 217–225 (2004)
9. Mewes, H.W., Frishman, D., Gldener, U., Mannhaupt, G., Mayer, K., Mokrejs, M., Morgenstern, B., Mnsterktter, M., Rudd, S., Weil, B.: MIPS: a database for genomes and protein sequences. Nucleic Acids Res. **30**(1), 31–34 (2002)
10. Mi, J.S., Leung, Y., Zhao, H.Y., Feng, T.: Generalized fuzzy rough sets determined by a triangular norm. Inf. Sci. **178**(16), 3203–3213 (2008)
11. Mitra, S., Pal, S.K.: Self-organizing neural network as a fuzzy classifier. IEEE Trans. Syst. Man Cybern. **24**(3), 385–399 (1994)
12. Pal, S.K., Dasgupta, B., Mitra, P.: Rough self-organizing map. Appl. Intell. **21**(1), 289–299 (2004)
13. Pal, S.K., Ghosh, A., Shankar, B.U.: Segmentation of remotely sensed images with fuzzy thresholding, and quantitative evaluation. Int. J. Remote Sens. **21**(11), 2269–2300 (2000)
14. Pal, S.K., Majumder, D.D.: Fuzzy sets and decision making approaches in vowel and speaker recognition. IEEE Trans. Syst. Man Cybern. **7**(8), 625–629 (1977)
15. Ray, S.S., Bandyopadhyay, S., Pal, S.K.: Dynamic range based distance measure for microarray expressions and a fast gene ordering algorithm. IEEE Trans. Syst. Man Cybern. Part B **37**, 742–749 (2007)
16. Sen, D., Pal, S.K.: Generalized rough sets, entropy and image ambiguity measures. IEEE Trans. Syst. Man Cybern. Part B **39**(1), 117–128 (2009)

Chapter 4
Fuzzy Rough Granular Neural Network and Unsupervised Feature Selection

4.1 Introduction

The problem of feature selection and its significance have been described in Sect. 1.5.2, Chap. 1. The objective of this task is to retain/generate the optimum salient characteristics necessary for the recognition process and to reduce the dimensionality of the measurement space so that effective and easily computable algorithms can be devised for efficient classification. For example, consider the classification and clustering algorithms in Chaps. 2 and 3. If a dataset contains redundant/irrelevant features, then that can degrade the performance and increase the computational complexity of clustering and classification algorithms.

The present chapter deals with the problems of feature selection in the framework of granular neural computing. Several algorithms for feature selection are available in [6, 9, 10, 12, 13]. Algorithms developed in neural networks framework can be found in [9, 10, 13]. These algorithms evaluate the importance of individual feature, based on a learning procedure, during training. In [13], a multi layer perceptron is used and trained with augmented cross entropy function whereas in [10], the formation of the network involves weighted distance between a pair of patterns, a membership function and a feature evaluation measure. This chapter is centered around a granular neural network [4], recently developed, for ordering and selecting important features in an unsupervised manner. While in the previous chapter, the crude domain knowledge about the data, as extracted using fuzzy rough sets in unsupervised mode, was encoded into the self organizing map for clustering, the same is used here in designing the granular neural network for feature selection.

We describe the aforesaid unsupervised granular neural network with their various characteristics and performance in Sect. 4.5. Before that a neural network based and a fuzzy neural network based feature selection algorithms are explained in Sects. 4.2 and 4.3, respectively. Section 4.4 describes some new notions of the lower and upper

© Springer International Publishing AG 2017
S.K. Pal et al., *Granular Neural Networks, Pattern Recognition and Bioinformatics*, Studies in Computational Intelligence 712,
DOI 10.1007/978-3-319-57115-7_4

approximations of a set, representing a fuzzy rough set, which have been used in the development of the method in Sect. 4.5.

4.2 Feature Selection with Neural Networks

A three layered neural network for feature selection [13] is described. The network is trained by minimizing the augmented cross-entropy error function based on gradient decent method. The cross-entropy error function, E, is defined as

$$E = \frac{E_0}{n_L} + \alpha_1 \frac{1}{P n_h} \sum_{p=1}^{P} \sum_{k=1}^{n_h} f'(net_{kp}^{(h)}) + \alpha_2 \frac{1}{P n_L} \sum_{p=1}^{P} \sum_{j=1}^{n_L} f'(net_{jp}^{(L)}), \qquad (4.1)$$

where α_1 and α_2 are parameters to be chosen experimentally, P is the number of training samples, n_L is the number of nodes in the output layer, $f'(net_{kp}^{(h)})$ $f'(net_{jp}^{(L)})$ are derivatives of the transfer functions of the kth hidden and jth output nodes, respectively, and

$$E_0 = -\frac{1}{2P}[\sum_{p=1}^{P} \sum_{j=1}^{n_L=Q} (d_{jp}log(o_{jp}^{(L)}) + (1 - d_{jp})log(1 - o_{jp}^{(L)}))], \qquad (4.2)$$

where d_{jp} is the desired output for the pth data point at the jth output node and Q is the number of classes. The feature selection procedure is summarized as following.

1. Initialize weights, randomly chosen small real numbers, to links connecting the nodes in three layered neural network (input, hidden and output layers). The architecture of the network is similar to that of multi layer neural network model.
2. Train the network, with training data, by minimizing the cross-entropy error function (using Eq. 4.1) based on gradient decent method. Update the weights of the network.
3. Arrange the updated connection weights between the input and hidden layers, representing the features, in decreasing order.
4. Rank the features according to the maximum updated weight values and calculate classification accuracy for test data with all the features.
5. Remove a feature with higher rank from all the features of the training and test data.
6. Repeat steps (2–5) using the training and test data with the selected features until there is no change in the classification accuracies of the test data.

It may be suggested to refer [13] for further details of the method. While this selection method, based on multilayer neural networks, is supervised, the one to be described in Sect. 4.3 is unsupervised.

4.3   Fuzzy Neural Network for Unsupervised Feature Selection

The formulation of a fuzzy neural network [10] is based on minimization of a feature evaluation index, where it incorporates membership function involving a weighted distance. The feature evaluation index is defined as

$$E = \frac{2}{s(s-1)} \sum_p \sum_{q \neq p} \frac{1}{2} [\mu_{pq}^T (1 - \mu_{pq}^O) + \mu_{pq}^O (1 - \mu_{pq}^T)]. \tag{4.3}$$

The characteristics of the index E are discussed in [10].

Computation of membership function: The membership function (μ) in feature space is defined as

$$\mu_{pq} = \begin{cases} 1 - \frac{d_{pq}}{D}, & \text{if } d_{pq} \leq D \\ 0, & \text{otherwise.} \end{cases} \tag{4.4}$$

d_{pq} is a distance measure which provides similarity (in terms of proximity) between pth and qth patterns in the feature space. Note that, the higher the value of d_{pq} is, the lower is the similarity between pth and qth patterns, and *vice versa*. D is a parameter which reflects the minimum separation between a pair of patterns belonging to two different clusters. The term D is expressed as

$$D = \beta d_{max}, \tag{4.5}$$

where d_{max} is the maximum separation between a pair of patterns in the entire feature space, and $0 < \beta \leq 1$.

The distance d_{pq} is defined as

$$d_{pq} = [\sum_i (x_{pi} - x_{qi})^2]^{1/2}, \tag{4.6}$$

where x_{pi} and x_{qi} are values of ith feature of pth and qth patterns, respectively. The d_{pq} is also defined as weighted distance as in [10].

Layered neural network: The network consists of an input, a hidden, and an output layer. The input layer consists of a pair of nodes corresponding to each feature, i.e., the number of nodes in the input layer is $2n$ for n-dimensional (original) feature space. The hidden layer consists of n number of nodes. The output layer consists of two nodes. One of them computes μ^O, and the other computes μ^T. The index E (using Eq. 4.4) is calculated from the μ-values of the network. Input nodes receive activations corresponding to feature values of each pair of patterns. A jth hidden node is connected only to an ith and $(i + n)$th input nodes via weights $+1$ and -1, respectively, where $j, i = 1, ..., n$ and $j = i$. The output node computing μ^T-values is connected to a jth hidden node via weight W_j, whereas that computing μ^O-values is connected to all the hidden nodes via weights $+1$ each.

During training, each pair of patterns are presented at the input layer and the evaluation index is computed. The weights W_j are updated using gradient-descent technique in order to minimize the index. Minimization of the index using the gradient-descent method with respect to the weights connecting the nodes in the hidden and output layers is described in [10].

4.4 Fuzzy Rough Set: Granulations and Approximations

In this section, a new fuzzy rough set is developed by defining new notions to lower and upper approximations of a set in the tolerance approximation space. Granulation of universe and approximation of a set are performed using the concepts of the fuzzy rough set.

As mentioned in Sect. 3.7.4, Chap. 3, decision system is represented by $SS = (U, \mathcal{A} \cup \{d\})$. Here U represents the universe, $\mathcal{A}$ represents conditional attributes (say, $\{a_1, a_2, \ldots, a_n\}$) and $\{d\}$ denotes a decision attribute. A fuzzy reflexive relation R_a, between any two patterns x and y in U, with respect to an attribute $a \in \mathcal{A}$ is defined in Eq. 2.43 (see Sect. 2.9.1, Chap. 2). For an attribute $a \in \{d\}$, the decision classes in fuzzy case are computed using Eq. 3.16. The lower and upper approximations of a set $A \subseteq U$ are described as follows.

4.4.1 New Notions of Lower and Upper Approximations

Here, we discuss the fuzzy logical operators involved in defining a fuzzy rough set. An operator T, mapping from $[0, 1]^2$ to $[0, 1]$, satisfies $T(1, x) = x$, for all $x \in [0, 1]$. Let T_M and T_P denote t-norms, and these are defined as

$$T_M(x, y) = min(x, y) \quad \text{(minimum operator) and,} \tag{4.7}$$

$$T_P(x, y) = x * y, \tag{4.8}$$

for all x and $y \in [0, 1]$. On the other hand, a mapping $I : [0, 1] \times [0, 1] \rightarrow [0, 1]$ satisfies $I(0, 0) = 1$, $I(1, x) = x$, for all $x \in [0, 1]$ where I is an implication operator. For all x and $y \in [0, 1]$, the implication operators I_{KD} and I_{KDL} are defined as

$$I_{KD}(x, y) = max(1 - x, y) \quad \text{(Kleene-Dienes implicator),} \tag{4.9}$$

$$I_{KDL}(x, y) = 1 - x + x * y \quad \text{(Kleene-Dienes-Lukasiewicz implicator).} \tag{4.10}$$

Lower Approximation: The membership value of pattern x in $A \subseteq U$ belonging to the lower approximation of a set A, based on a fuzzy $\mathcal{B} \subseteq \mathcal{A}$-indiscernibility relation, fuzzy decision classes R_d, and fuzzy logic connectives, (Eqs. 4.8 and 4.9), is defined as

$$(R_B \downarrow R_d)(x) = \min\{\underline{\gamma}(x), \underline{\gamma}^c(x)\}, \tag{4.11}$$

where

$$\underline{\gamma}(x) = \inf_{y \in A}\{R_B(x, y) * R_d(x, y)\}, \tag{4.12}$$

$$\underline{\gamma}^c(x) = \inf_{y \in U - A}\{\max(1 - R_B(x, y), R_d(x, y))\}. \tag{4.13}$$

Here, $\underline{\gamma}(x)$ represents a membership value of a pattern x, computed by the weighted product of $R_B(x, y)$ and $R_d(x, y)$, when the patterns x and y belong to a particular set (class) A. While $R_B(x, y)$ represents a membership value of the pattern in the fuzzy reflexive relational matrix with respect to an attribute a in $\mathcal{B}$, $R_d(x, y)$ refers to a membership value of the pattern belonging to the fuzzy decision class. The value of $\gamma^c(x)$ lying in the interval $[0, 1]$ is computed (using Eq. 4.9), when the patterns x and y belong to two different sets (classes).

Upper Approximation: The membership value of a pattern x in $A \subseteq U$ for belonging to upper approximation of a set A, based on the fuzzy reflexive relation R_B, and fuzzy logic connectives, (see Eqs. 4.7 and 4.10), is defined as

$$(R_B \uparrow R_d)(x) = \max\{\overline{\gamma}(x), \overline{\gamma}^c(x)\}, \tag{4.14}$$

where

$$\overline{\gamma}(x) = \sup_{y \in A}\{1 - R_B(x, y) + (R_B(x, y) * R_d(x, y))\} \text{ and} \tag{4.15}$$

$$\overline{\gamma}^c(x) = \sup_{y \in U - A}\{\min(R_B(x, y), R_d(x, y))\}, \tag{4.16}$$

for all x in U. Here, the term $\overline{\gamma}(x)$ represents a membership value, (computed by using Eq. 4.10), when the two patterns x and y belong to a set A. The value of $\overline{\gamma}^c(x)$ lies between $[0, 1]$ and is computed (using Eq. 4.7), when two patterns belong to two different sets (not in the same set A).

The fuzzy positive region, based on the fuzzy $\mathcal{B}$-indiscernibility relation, is defined using Eq. 2.36. The dependency values, γ_B, of a set of attributes, $\mathcal{B} \subseteq \mathcal{A}$, are calculated using Eq. 2.38 (see Sect. 2.8.1, Chap. 2).

The notions of the lower and upper approximations of a set are explained in details in the following steps.

1. Normalize the given data meaning that the values of features (attributes) lie within 0 to 1.

2. Compute reflexive relational matrix of the size $s \times s$ for every feature using Eq. 2.43. There are s rows and s columns in the matrix.
3. Find fuzzy decision attribute values, representing fuzzy decision classes, corresponding to every pattern in universe using Eq. 3.16. The number of decision attributes for every pattern is equal to the number of classes. This process involves the following points

   (a) Compute mean of each of the classes using the decision table SS.
   (b) Calculate distance from every pattern to mean of its own class.
   (c) Find membership value for pattern belonging to its own classes by normalizing the distance value using fuzzifiers.
   (d) Define decision attribute corresponding every pattern with the membership for its actual class and 1 for the other classes.

4. Find the notions of the lower and upper approximations of a set using Eqs. 4.11 and 4.14, respectively, based on the aforesaid fuzzy relational matrices and fuzzy decision classes.

4.4.2 Scatter Plots of Features in Terms of Lower and Upper Approximations

Let us consider a data set containing two classes shown in Table 4.1, as a typical example. Note that, the table is already shown in Sect. 2.9.2 and is again provided for the convenience. Here, the data is normalized within 0 to 1. The membership values of the patterns belonging to lower and upper approximations, corresponding to every feature, are defined using Eqs. 4.11 and 4.14, respectively. Figure 4.1 shows a 3-dimensional scattered plot of fuzzy attribute values belonging to lower and upper approximations in F_1-F_2-F_3 space. It can be observed from the figure that the values of patterns belonging to the lower approximations of the two classes (sets) are lower than those belonging to upper approximations of the two classes.

Table 4.1 Dataset

U	a	b	c	d
x_1	−0.4	−0.3	−0.5	1
x_2	−0.4	0.2	−0.1	2
x_3	−0.3	−0.4	0.3	1
x_4	0.3	−0.3	0	2
x_5	0.2	−0.3	0	2
x_6	0.2	0	0	1

Fig. 4.1 3-dimensional
scattered features of the data,
defined in terms membership
values for belonging to lower
and upper approximations, in
the F_1-F_2-F_3 space

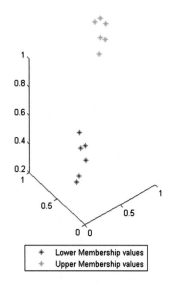

4.5 Fuzzy Rough Granular Neural Network for Unsupervised Feature Selection

4.5.1 Strategy

The problem of ranking the features by designing a granular neural network [4] using the concepts of fuzzy sets and fuzzy rough sets is described. It involves notion of a fuzzy rough set, characterized by lower and upper approximations of a set, and a three-layered fuzzy rough granular network (FRGNN3) for feature selection. The construction of the network is based on input and target value, fuzzy initial connection weights and feature evaluation index which are defined using granulation structures based on α-cut. The feature evaluation index involves the error value at output node of the network where the error is obtained by calculating the distance between target value and the output value (obtained value at the output node). The output value is weighted product of initial connection weights and input vector. The network is trained through minimization of the feature evaluation index using gradient decent method. The updated weights reflect the relative importance of feature in measuring the intra cluster distance and inter cluster distance. As the error value decreases, the value of feature evaluation index becomes minimum. This implies that, if the feature evaluation index corresponding to the set of selected features decreases, the intra cluster distances decrease and inter cluster distances increase. Note that, this network performs feature selection in unsupervised manner and it needs to partition the features space into a particular number of clusters to fix the number of its input nodes. But the network is not dependent particularly on class information available with the data.

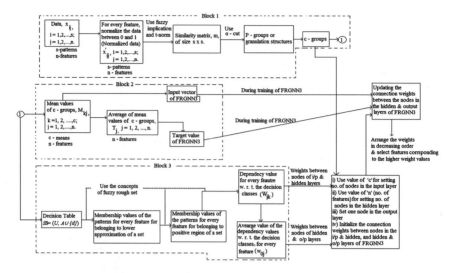

Fig. 4.2 Diagrammatic representation of the FRGNN3

A diagrammatic representation of the fuzzy rough granular neural network (FRGNN3) for unsupervised feature selection is provided in Fig. 4.2. The main steps are divided into three blocks, shown with dotted lines. These are as follows:

(1) *Generate c-granulation structures*: The first block within dotted lines (Block 1) represents the procedure for determining the c-granulation structures which includes data normalization, formation of a similarity matrix using fuzzy implication and t-norm operators, generation of p-granulation structures (groups) and selection of top c-granulation structures from them. These are explained in Sects. 4.5.2 and 4.5.3.

(2) *Determine the input vector and target value of FRGNN3*: The second block (Block 2) provides an approach for determining the input vector and target value of FRGNN3 where the input vector is represented by the mean values of the c-granulation structures, and the target value is defined by average of the mean values, corresponding to a feature. This approach is explained in Sect. 4.5.4.

(3) *Determine initial connection weights of FRGNN3*: The procedure for determining the connection weights of FRGNN3, including membership values of the patterns belonging to lower approximation and positive region of a set corresponding to the features, dependency values for the features with respect to the decision classes, and average values of the dependency values for the features (using the concepts of the fuzzy rough set as explained in Sect. 4.4) is provided in the third block (Block 3) and is described in Sect. 4.5.5.

Now, the output of Blocks 1 and 3 are used for setting the number of nodes in the input layer (c) and hidden layer (n) and initializing the connection weights between the nodes in the input and hidden and hidden and output layers of FRGNN3, respectively. Block 2 provides the input vector and target value for training FRGNN3.

The architecture of FRGNN3 is then formulated and is explained in the first paragraph of Sect. 4.5.5.

The mechanism involved in the forward propagation and minimization of the feature evaluation index with respect to the connection weights between nodes of the hidden and output layers are explained in Sect. 4.5.5.

4.5.2 Normalization of Features

Let $\{x_{ij}\}$, $i = 1, 2, \ldots, s$; $j = 1, 2, \ldots, n$; be a set of n-dimensional data where s represents the total number of patterns in the data. The data is then normalized between 0 and 1, for every feature. For $j = 1$, the data is x_{i1}. The procedure for normalization of x_{i1} is defined as

$$x'_{i1} = \frac{x_{i1} - x_{min_{i1}}}{x_{max_{i1}} - x_{min_{i1}}}, \tag{4.17}$$

where $x_{min_{i1}}$ and $x_{max_{i1}}$ are the minimum and maximum values, respectively, of a feature x_{i1} for $j = 1$. In a similar way, normalization is performed for $j = 2, \ldots, n$. The resultant data is then used to determine granulation structures based on α-cut.

4.5.3 Granulation Structures Based on α-Cut

The data is partitioned into granulation structures by applying a user defined α-cut on a pairwise similarity matrix computed using normalized data, x'_{ij}. For example, the steps for computing the similarity value, say m_{12}, between x'_{1j} and x'_{2j}, using implication and t-norms (Eqs. 2.24 and 2.26, Sect. 2.8 in Chap. 2) are shown as follows:

1. $I_{1j} \leftarrow \min(1 - x'_{1j} + x'_{2j}, 1)$, $I_{2j} \leftarrow \min(1 - x'_{2j} + x'_{1j}, 1)$,
2. $T_j \leftarrow \max(I_{1j} + I_{2j} - 1, 0)$,
3. $m_{12} \leftarrow \min \{T_j\}$.

In a similar manner, we find the similarity values between all possible pairs of patterns and construct a similarity matrix, m, of size $s \times s$ where s is the total number of patterns. The procedure of generating the granulation structures is already explained (see the algorithms in Sect. 3.7.3, Chap. 3). By using the similarity matrix m, we generate p number of granulation structures/groups, based on an α-cut, chosen between 0 and 1. The steps are as follows:

S1: Let i and i' denote the number of rows and columns of the similarity matrix m where i and $i' = 1, 2, \ldots, s$.
S2: Choose an α value between 0 and 1.

S3: Apply the α-value or a threshold on the first row ($i = 1$) of the similarity matrix m, and find how many similarity values in that row are greater than the α-value.

S4: Find the positions of the columns, corresponding to the similarity values greater than the α-value, in the first row of the similarity matrix.

S5: Select the patterns, corresponding to the positions of the columns in the first row of the similarity matrix, to form one group.

S6: Skip the rows and also columns corresponding to the selected positions in the similarity matrix of Step S4.

S7: Apply steps S4–S6 on the remaining rows in the similarity matrix, until all the patterns are assigned into the groups.

The granulation structures (p groups) generated using the above mentioned procedure, are arranged in decreasing order according to their sizes where the size is defined by the number of patterns within a group. We choose the top c groups, out of the p groups, for formation of the FRGNN3 where c is a user defined value. The data points in $c + 1$ to p groups are then added into the top c-groups, as described in the following Steps:

(1) Find average values (means) for all p-groups (including the top c-groups and ($c + 1$ to p)).

(2) Find the Euclidean distances from the means of top c-groups to the mean of the $c + 1$ group.

(3) Find the minimum Euclidean distance and add all the patterns in $c + 1$ group into the group corresponding to the minimum distance.

(4) Repeat Steps 2 and 3 for $c + 2$ to p-groups, until all the patterns in these groups are added into the top c-groups.

The resultant c-groups (granulation structures) are then used in the formation of the FRGNN3.

4.5.4 Determination of Input Vector and Target Values

Let $\{x_{ij}^{k}\}$ denote a set of n-dimensional normalized patterns belonging to the c-granulation structures where $k = 1, 2, \ldots, c; i = 1, 2, \ldots, s_k; j = 1, 2, \ldots, n$. Here, s_k is the number of patterns in the kth granulation structure. The average values of the patterns in c-granulation structures are calculated and a mean matrix, denoted by M_{kj}, of the size $c \times n$, is constructed where rows represent the means of the groups and columns represent the features. Let $\overrightarrow{I}^{\,j}$ and TG_j denote an input vector and a target value corresponding to a feature j. $\overrightarrow{I}^{\,j}$ is defined as

$$\overrightarrow{I}^{\,j} \equiv \{M_{kj}\}, k = 1, 2, \ldots, c. \tag{4.18}$$

$\overrightarrow{I}^{\,j}$ represents an input vector corresponding to the jth column in the mean matrix M_{kj}. TG_j is defined as

Table 4.2 Input vector and target value of the FRGNN3

	Mean of each of the c-groups				
	Feature f_1	Feature f_2	Feature f_3	$\ldots$	Feature f_n
Group$_1$	M_{11}	M_{12}	M_{13}	$\ldots$	M_{1n}
Group$_2$	M_{21}	M_{22}	M_{23}	$\ldots$	M_{2n}
Group$_3$	M_{31}	M_{32}	M_{33}	$\ldots$	M_{3n}
$\vdots$	$\vdots$	$\vdots$	$\vdots$	$\vdots$	
Group$_c$	M_{c1}	M_{c2}	M_{c3}	$\ldots$	M_{cn}
	Average of data (Average of the means)				
Data in groups $1, \ldots, c$	TG_1	TG_2	TG_3	$\ldots$	TG_n

$$TG_j = \frac{\sum_{k=1}^{c} M_{kj}}{c}. \tag{4.19}$$

In other words, TG_j represents a target value computed by the average of the components of the input vector corresponding to the jth feature. Here, the component is the mean of the granulation structure along the feature. We explain concisely the concept of input vector and target value by the mean matrix shown in Table 4.2.

We can see from Table 4.2 that a mean matrix of the size $c \times n$ where c is the number of groups and n is the number of features. For example, considering feature f_1, the mean values of all groups, $\{M_{11}, M_{21}, M_{31}, \ldots, M_{c1}\}$, and the average of the means, TG_1, are taken as input vector and target value, respectively, of FRGNN3.

4.5.5 Formation of the Fuzzy Rough Granular Neural Network

We consider a three layer neural network consisting of interconnected neurons (nodes) for formation of FRGNN3. The number of nodes in the input layer is set equal to c as there are c-rows in the mean matrix M_{kj}. In the hidden layer, the number of nodes is set equal to the number of features (n). One node is set in the output layer. The links connecting the nodes in the input layer and hidden layer, and hidden layer and output layer are initialized by the connection weights, say W_{jk} and w_{0j}, respectively. The initial architecture of FRGNN3 is shown in Fig. 4.3.

4.5.5.1 Initial Connection Weights

The concepts of fuzzy rough set, based on a decision table, are used for extracting the domain knowledge about data. It is incorporated in FRGNN3 as its initial connection

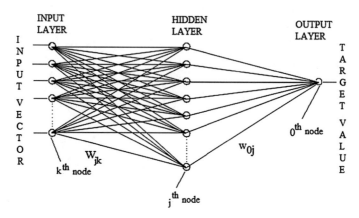

Fig. 4.3 FRGNN3 architecture

weights W_{jk} and w_{0j}. The procedure for defining the connection weights is described as follows:

Procedure: The c-granulation structures obtained by an α-cut are presented to the aforesaid decision system $SS = (U, \mathcal{A} \cup \{d\})$, where every structure is labeled with a value representing the decision classes. The following steps are applied to the decision table SS for extracting domain knowledge about data.

S1: Generate a fuzzy reflexive relational matrix for every feature by computing fuzzy relations between all possible pairs of patterns. The fuzzy relations in each row of the matrix constitute a fuzzy equivalence granule.

S2: For each conditional attribute, compute the membership value (using Eq. 4.11 or Eq. 2.34), belonging to the lower approximation of every pattern in a concept using fuzzy reflexive relational matrix and fuzzy decision classes.

S3: Calculate the fuzzy positive region (using Eq. 2.36) of every pattern for each feature.

S4: Compute the dependency value (using Eq. 2.38) of each feature with respect to concept (class). Assign these dependency factors as initial connection weights, denoted by W_{jk}, between nodes of the input layer and the hidden layer of FRGNN3.

S5: Find the average dependency degree of all the dependency value of features with respect to the concepts. Here, the average dependency degrees can be represented as the dependency value of the feature. Assign the resultant dependency values of the conditional attributes as initial connection weights, denoted by w_{0j}, between nodes of hidden and output layers of FRGNN3.

FRGNN3 is then trained, in presence of input vector and target vector, using forward propagation and minimization of feature evaluation index with respect to the connection weights between the nodes in the hidden layer and output layer (w_{0j}), in an unsupervised manner. The connection weights w_{0j} are updated, as these weights (w_{0j}) represent dependency values of the features (see Step S5). The updated weights provide the ordering or importance of the individual features. A higher dependency degree of the attribute signifies that the attribute/feature is better for selection. It may be noted that the connection weights between the nodes in the input layer and hidden layer (W_{jk}) are not updated during training.

4.5.5.2 Algorithm for Training

We first explain forward propagation of FRGNN3. A feature evaluation index is then defined using an error value, attained at the node of output layer, after completing the forward propagation. The feature evaluation index with respect to the connection weights between the nodes in the hidden and output layers is minimized using gradient decent method. Based on the minimization process the connection weights (w_{0j}) are modified during training. An algorithm for training FRGNN3 is provided at end of this Section.

Forward propagation: During forward propagation, an input vector $\overrightarrow{I}^{\,j}$ (Eq. 4.18) corresponding to a feature j is presented at the input layer.

Assume that W_{jk} (see Step 4) is a initial connection weight from kth node in the input layer to jth node in the hidden layer. Here, W_{jk} represents the dependency factor of jth feature, corresponding to the kth decision class. Therefore, the input of the jth hidden node is the sum of outputs of k input nodes received via connection weights W_{jk} and is represented by

$$O_j^{(1)} = \sum_{k=1}^{c} W_{jk} I_k^j, \ 1 \le j \le n, \tag{4.20}$$

The total output of jth hidden node is the input for single node 0 in the output layer. Let w_{0j} be the initial connection weight (see Step 5) from jth node in the hidden layer to single node 0 in the output layer. Therefore, the sum of activations received via connection weights w_{0j} by the single node in the output layer is represented as

$$O^{(2)} = \sum_{j=1}^{n} w_{0j} O_j^{(1)}, \tag{4.21}$$

After the forward propagation of FRGNN3 is completed, we calculate error at the output layer node, denoted by E, using the following equation:

$$E = \eta (TG - O^{(2)})^2, \tag{4.22}$$

where $O^{(2)}$ is the obtained output or actual output at the node in output layer, T is the target value and η is a parameter chosen in the interval $(0, 1]$ and is used to put the value of E within 0 to 1, while an input vector is presented at the nodes in the input layer of FRGNN3. Here, the error is defined by calculating the Euclidean distance

Fig. 4.4 Feature evaluation index for different values of E

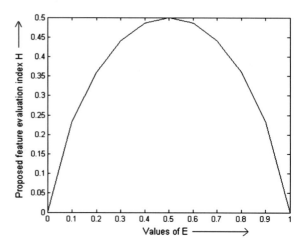

between the target value and the obtained output. A fuzzy feature evaluation index is then given in terms of E as follows:

Feature evaluation index: A fuzzy feature evaluation index, denoted by H, is defined in terms of the error E (Eq. 4.22) as

$$H = \frac{1}{cln2}[-Eln(E) - (1 - E)ln(1 - E)], \tag{4.23}$$

where c is a constant and is set equal to the number of input nodes in FRGNN3 (which is equal to the number of granulation structures). Figure 4.4 shows the plot of the fuzzy feature evaluation index H with $c = 2$ and base e for different values of E $\in [0, 1]$ in 2D plane. The properties of the feature evaluation index H are discussed as follows:

1. Sharpness: $H = 0$ iff $E = 0$ or 1.
2. Maximality: H attains maximum value iff $E = 0.5$.
3. Symmetric: $H = \overline{H}$ iff $E = \overline{E}$ where $\overline{H}$, compliment of H, attains a value corresponding to a value of $\overline{E}$, complement of E.
4. Resolution: $H \geq H^*$, when $E \geq E^*$. Here, H^* varies with E^*, sharpened version of E.
5. Continuity: H is continuous function of E where $E \in [0, 1]$.

The feature evaluation index H is now minimized with respect to the weights, w_{0j}, between nodes of the hidden and output layers. The minimization of feature evaluation index H, with respect to the weights, signifies that the convergence towards the improved values for the weights is achieved. The improved values for the weights provide an ordering of the individual features. A higher value of the weight signifies that the importance of the corresponding feature is high.

Minimization of feature evaluation index: The minimization of feature evaluation index H with respect to the weights w_{0j} is performed by using gradient decent method. This implies that the changes in the weights w_{0j}, say $\triangle w_{0j}$, are proportional to $-\frac{\partial H}{\partial w_{0j}}$.

$$\triangle w_{0j} = -\delta \frac{\partial H}{\partial w_{0j}}, \tag{4.24}$$

where δ is the learning rate chosen between 0 and 1. The partial derivative $\frac{\partial H}{\partial w_{0j}}$ can be evaluated using the chain rule

$$\frac{\partial H}{\partial w_{0j}} = \frac{\partial H}{\partial O^{(2)}} \frac{\partial O^{(2)}}{\partial w_{0j}}. \tag{4.25}$$

Using Eq. 4.21, the expression $\frac{\partial O^{(2)}}{\partial w_{0j}}$ can be obtained as

$$\frac{\partial O^{(2)}}{\partial w_{0j}} = \frac{\partial}{\partial w_{0j}} \sum w_{0j} O_j^{(1)} = O_j^{(1)}. \tag{4.26}$$

The expression $\frac{\partial H}{\partial O^{(2)}}$ signifies the rate of change of error with respect to the output $O^{(2)}$ and using Eq. 4.23 it can be written as

$$\frac{\partial H}{\partial O^{(2)}} = \frac{\partial}{\partial O^{(2)}} (\frac{1}{cln2} [-Eln(E) - (1 - E)ln(1 - E)]). \tag{4.27}$$

Using Eq. 4.22, Eq. 4.27 can be written as

$$\frac{\partial H}{\partial O^{(2)}} = \frac{\partial}{\partial O^{(2)}} (\frac{1}{cln2} \{-(\eta(T - O^{(2)})^2)ln(\eta(T - O^{(2)})^2) -(1 - \eta(T - O^{(2)})^2)ln(1 - \eta(T - O^{(2)})^2)\}). \tag{4.28}$$

After completion of the differentiation, Eq. 4.28 can be obtained as

$$\frac{\partial H}{\partial O^{(2)}} = \frac{1}{cln2} (2\eta(T - O^{(2)})ln(\eta(T - O^{(2)})) -(2\eta(T - O^{(2)})ln(1 - \eta(T - O^{(2)})^2))). \tag{4.29}$$

Now, we substitute Eqs. 4.26 and 4.29 in Eq. 4.25. Therefore, Eq. 4.25 is obtained as

$$\frac{\partial H}{\partial w_{0j}} = \frac{1}{cln2} (2\eta(T - O^{(2)})ln(\eta(T - O^{(2)})) -(2\eta(T - O^{(2)})ln(1 - \eta(T - O^{(2)})^2)))O_j^{(1)}. \tag{4.30}$$

Using Eqs. 4.24 and 4.30, we obtain,

$$\triangle w_{0j} = -\delta(\frac{1}{cln2} (2\eta(T - O^{(2)})ln(\eta(T - O^{(2)})^2) -(2\eta(T - O^{(2)})ln(1 - \eta(T - O^{(2)})^2)))O_j^{(1)}). \tag{4.31}$$

During training of FRGNN3, the weights w_{0j} of the links, connecting the hidden and the output layers, are updated, and the updated equation is defined as

$$w_{0j}(e+1)+ = \triangle w_{0j}(e); \qquad (4.32)$$

where e represents an epoch/iteration.

Steps used for training the network: For every iteration, the training of the network is performed using the following steps:

S1: Present an input vector $\overrightarrow{I}_j$ (Eq. 4.18) at the nodes of the input layer and a target vector TG_j (Eq. 4.19) at the output layer node.

S2: Train the network using Eqs. 4.20 and 4.21.

S3: Find $\triangle w_{0j}$ using Eq. 4.31.

S4: Update the weights w_{0j} using Eq. 4.32.

After the training of FRGNN3 is completed, the weights w_{0j} are arranged in decreasing order. A higher value (closer to 1) of w_{0j} signifies that the corresponding feature j is better for selection. The number of features in a set is represented by Z which is less than or equal to the total number of features. The top-Z number of selected features are used in experiment.

The performance of FRGNN3 is compared with unsupervised feature selection using feature similarity (UFSFS [6]), sequential forward search (SFS [2]) and Relief-F [5]. The features selected by the methods are evaluated using Naive Bayes and K-NN and the entropy measure for feature evaluation (feature evaluation index [8]).

In addition, a self-organizing map is used for clustering the microarray gene expression data, based on the selected features. The resultant clusters of microarray gene expression data sets are evaluated using the entropy, β-index [11], Davies-Bouldin index (DB-index) [1] and fuzzy rough entropy [3]. A higher value of β-index indicates that the clustering solutions are compact whereas it is the opposite for entropy, DB-index and fuzzy rough entropy.

4.6 Experimental Results

In this section, the performance of FRGNN3 is tested on different real life data sets, including microarray gene expression data. The characteristics of the real-life data sets, collected from the UCI machine learning repository [7], are shown in Table 4.3.

Data sets are chosen in three categories based on their dimensions such as *low* (dimension < 10), *medium* (10 < dimension < 100) and *high* (dimension > 100). The data sets, iris, Wisconsin cancer, waveform and multiple features, are summarized in Appendix. Microarray gene expressions, like Cell Cycle, Yeast Complex, All Yeast data, are also used (see Sect. 3.9, Chap. 3 for explanation).

Selection of features: We first explain the connectionist mechanism of FRGNN3 and then describe the process of selecting the best features by using FRGNN3 for

Table 4.3 Characteristics of data

Data set name	No. of patterns	No. of features	No. of classes
Iris	150	4	3
Wisconsin cancer	684	9	2
Waveform	5000	40	3
Multiple features	2000	649	10

Fig. 4.5 F_1-F_2 plane

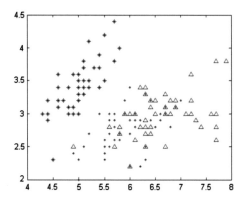

Fig. 4.6 F_1-F_3 plane

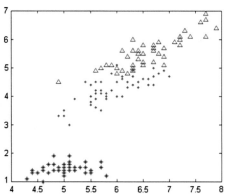

iris data. Let F_1, F_2, F_3 and F_4 denote the four consecutive features of iris data (sepal length, sepal width, petal length, petal width). Figures 4.5, 4.6, 4.7, 4.8, 4.9 and 4.10 show the scatter plots of the pairwise combinations of the four features of three classes in the 2D plane.

Initially, each feature is normalized within 0 and 1 using Eq. 4.17 (see Sect. 4.5.2). A similarity matrix is computed as explained in Sect. 4.5.3. Different numbers of granulation structures are then generated for different values of α, chosen between 0 and 1. As a typical example, let us consider the iris data. Here, 11 groups are generated for $\alpha = 0.69$. These groups are then arranged in a descending order according to their sizes. We choose the top 5 groups ($c = 5$) out of 11 groups. The remaining

Fig. 4.7 F_1-F_4 plane

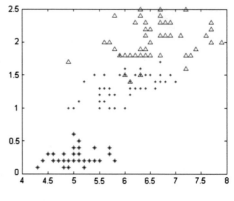

Fig. 4.8 F_2-F_3 plane

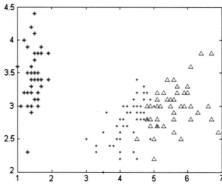

Fig. 4.9 F_2-F_4 plane

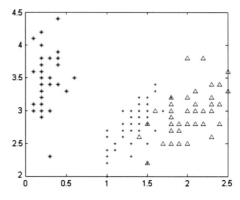

Fig. 4.10 F_3-F_4 plane

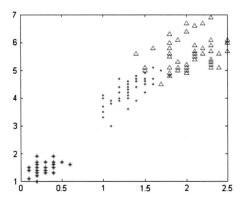

Table 4.4 A mean matrix, representing input vector, defined by using the 5 groups for iris data

	F_1	F_2	F_3	F_4
Input vector	0.3908	0.6048	0.3033	0.3376
	0.8049	0.4036	0.9212	0.9409
	0.6355	0.7533	0.4583	0.485
	0.3680	0.4930	0.2274	0.0520
	0.5833	0.6726	0.5544	0.5416
Target	0.5712	0.5594	0.5325	0.5419

groups (6 to 11) are added into the top 5-groups using the Steps 1–5, in Sect. 4.5.3. The input vectors and the target values are then determined using Eqs. 4.18 and 4.19, respectively, and are shown in Table 4.4. By observing the results from Table 4.4, an input vector and target value corresponding to feature F_1 is represented by {0.3908, 0.8049, 0.6355, 0.3680 and 0.5833} and 0.5712, respectively. In FRGNN3, the number of nodes in the input layer is set to 5 for iris data as the mean matrix has 5 means corresponding to the 5 groups. The number of nodes in the hidden layer is set to 4 as the iris data has four features. The 5 groups are presented to a decision system SS. The decision table SS is used to extract domain knowledge about data in terms of dependency factors using the concept of fuzzy rough sets. The dependency factors of the features, corresponding to the 5 groups, and the average of them are initialized as the connection weights between nodes of the input layer and hidden layer, and the hidden layer and output layer, respectively. These are shown in Table 4.5.

During training of FRGNN3, the input vector is presented at nodes of the input layer, while the target value is stored at the single node of the output layer. The connection weights between nodes of hidden and output layers are updated using Eq. 4.32. The values of η (a parameter used to put the error within 0 to 1) in Eq. 4.22 and δ (learning rate) in Eq. 4.32 are set to be 0.9 and 0.00091, respectively, for iris data. After the training of FRGNN3 is completed, the features are ranked based on the updated weights (w_{0j}) between nodes of the hidden and output layers, as shown in Table 4.6, where the feature with maximum updated weight is assigned to top rank.

Table 4.5 Initial connection weights of FRGNN3 for iris data

					Initial connection weights from 4 nodes in the hidden layer (HN_1, HN_2, HN_3 and HN_4) to the node in the output layer (ON_1)	
Initial connection weights from five nodes (IN_1, IN_2, IN_3, IN_4 and IN_5) in the input layer to 4 nodes (HN_1, HN_2, HN_3 and HN_4) in the hidden layer						
	HN_1	HN_2	HN_3	HN_4		ON_1
IN_1	0.1010	0.1057	0.1197	0.1247	HN_1	0.1271
IN_2	0.1584	0.1495	0.2003	0.1806	HN_2	0.1513
IN_3	0.1711	0.1541	0.1783	0.1531	HN_3	0.1492
IN_4	0.1512	0.1617	0.1829	0.1956	HN_4	0.1317
IN_5	0.0767	0.0643	0.0754	0.0921		

Table 4.6 FRGNN3 ranks the features of iris data

Features	Updated weights	Order
F_1	0.3457	4
F_2	0.3563	3
F_3	0.4028	1
F_4	0.3732	2

The ordering is found to be $F_3 > F_4 > F_1 > F_2$. For evaluating the effectiveness of FRGNN3, we selected the top Z number of the features, say $Z = 2$, i.e., features F_3 and F_4, and classified the data set separately using Naive Bayes and K-NN classifiers, and the results are shown in terms of percentage of accuracies. The importance of the features are also evaluated with a entropy measure where for a particular method, say FRGNN3, the entropy value for the top Z number of selected features is calculated. The results are provided in Table 4.7. The same procedure, explained so far for iris data, is applied for selecting salient features from the remaining data sets.

Affect of top-Z features on classification accuracies for K-NN and Naive Bayes: The affect of top-Z features on classification accuracies for K-NN and Naive Bayes is described for two typical data sets, waveform and multiple features data with dimensions 40 and 649, respectively. Here, we provide the variation of classification accuracies, using K-NN and Naive Bayes classifiers, for different numbers of top-Z selected features, selected by FRGNN3, UFSFS, SFS and Relief-F.

Waveform data: For waveform data, the variation of the average classification accuracies with the numbers of selected features, obtained by all the algorithms, are shown in Figs. 4.11 and 4.12. It is observed that the classification accuracies of both K-NN and Naive Bayes for all the algorithms are gradually increased for $Z = 5$, 10 and 15, and are gradually decreased for $Z = 20, 25, 30$ and 35.

For $Z = 5$ and 10, the performance of FRGNN3 is inferior to Relief-F and is superior to UFSFS and SFS. For the remaining values of Z (other than $Z = 5$ and 10), it is evident from Figs. 4.11 and 4.12 that the performance of FRGNN3 is better

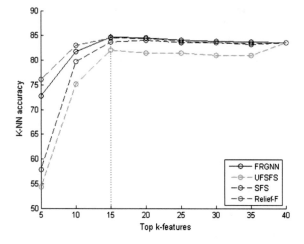

Fig. 4.11 Variation of the average classification accuracies of K-NN with the number of top selected features for waveform data

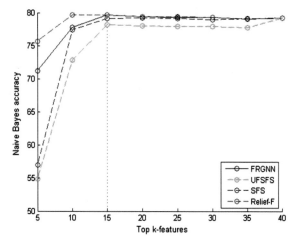

Fig. 4.12 Variation of the average classification accuracies of Naive Bayes with the number of top selected features for waveform data

than the related methods. At value 15, Z is marked with a vertical dotted line where the maximum performance is achieved for all algorithms.

Multiple features data: For multiple features data, the values of Z are chosen as 65, 130, 195, 260, 325, 390, 455, 520, 585 and 649, where the smallest one ($Z = 65$) is 10% of the actual feature size, and the other values are obtained by increasing the percentage by 10 at a time. Thereafter, the data with the selected number of features, obtained by FRGNN3, UFSFS, SFS, Relief-F algorithms, is classified using K-NN and Naive Bayes. The plots of average classification values of K-NN and Naive Bayes, corresponding to different values of Z, are shown in Fig. 4.13 and 4.14, respectively.

From Fig. 4.13, it can be seen that FRGNN3 provides higher classification accuracies, as compared with UFSFS, SFS, Relief-F, for $Z = 65$, 130, 195, 260, 325 and 390 using K-NN. Using Naive Bayes, similar observation can also be made for

Fig. 4.13 Variation of the
average classification
accuracies of K-NN with the
number of top selected
features for multiple features
data

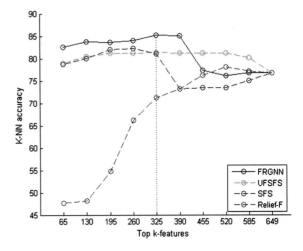

Fig. 4.14 Variation of the
average classification
accuracies of Naive Bayes
with the number of top
selected features for multiple
features data

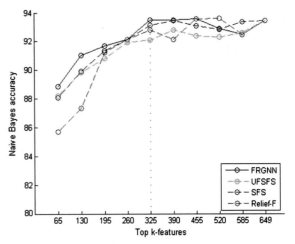

all the algorithms from Fig. 4.14 for $Z = 65$, 130, 195, 325 and 390. At value 260, the performance of FRGNN3 seems to be equal to SFS and Relief-F. However, the best classification accuracy, using K-NN and Naive Bayes, for all the algorithms is achieved at value 325, out of all other values Z. Hence, this value is marked with a vertical dotted line in Figs. 4.13 and 4.14.

The results for a particular value of Z, for which K-NN and Naive Bayes provide the best average performances (average of 10 fold cross validation results) for all algorithms, are shown in Table 4.7. The value of Z is mentioned within parenthesis at column 1. The CPU time for selection of features, classification accuracies and entropy value with the selected features are shown in the table for all the methods and data sets. The number of nodes set in the input layer of FRGNN3 is shown within parenthesis at column 2.

Table 4.7 Comparison of the classification values of K-NN and Naive Bayes for Relief-F, SFS, UFSFS and FRGNN3 for different data sets

Data set	Method	K-NN			Naive Bayes			Entropy	CPU time in sec.
		Min. Acc. (%)	Max. Acc. (%)	Avg. Acc. (%)	Min. Acc. (%)	Max. Acc. (%)	Avg. Acc. (%)		
Iris (Z = 2)	Relief-F	87.41	96.29	93.04	86.67	97.03	94.00	0.2078	0.4595
	SFS	62.96	91.11	82.37	60.71	93.93	81.87	0.2191	0.4056
	UFSFS	87.41	96.29	93.04	86.67	97.03	94.00	0.2078	0.0468
	FRGNN3(5)	87.41	96.29	93.04	86.67	97.03	94.00	0.2078	0.5000
Wisconsin Breast Cancer (Z = 6)	Relief-F	94.28	96.57	95.79	95.09	96.24	95.44	0.2158	1.1112
	SFS	94.77	96.41	95.75	95.42	97.05	96.11	0.2169	0.6252
	UFSFS	94.42	96.73	95.62	94.77	97.05	96.04	0.2158	0.0506
	FRGNN3(2)	95.42	96.57	96.06	95.75	97.05	96.45	0.2130	2.2560
Waveform (Z = 15)	Relief-F	83.68	85.37	84.58	78.71	80.75	79.42	0.2424	26.4391
	SFS	82.97	84.86	83.80	78.66	80.22	79.19	0.2426	20.3433
	UFSFS	81.55	82.35	82.06	77.13	79.88	78.23	0.2433	0.6978
	FRGNN3(3)	83.60	85.51	84.62	79.26	80.77	79.62	0.2418	146.0420
Multiple features (Z = 325)	Relief-F	68.88	74.94	70.70	91.27	94.66	92.78	0.2368	327.9212
	SFS	67.56	72.22	70.22	91.66	94.05	93.27	0.2475	19477.9917
	UFSFS	78.72	84.44	81.35	91.44	94.00	92.55	0.2465	63.5080
	FRGNN3(8)	81.05	87.11	85.15	91.94	94.66	93.42	0.2454	850.9470

For iris data, features 3 and 4 are selected using FRGNN3. The same features are also selected by UFSFS and Relief-F, except SFS where it selects features 1 and 3. For the data with the selected features, the classification accuracies from Table 4.7 for FRGNN3, as shown in terms of K-NN and Naive Bayes, and entropy value, are found to be equal to those of UFSFS and Relief-F. These are superior to SFS. Moreover, the average classification accuracies for FRGNN3 are superior to SFS.

The results from Table 4.7 for Wisconsin breast cancer, waveform and multiple features show that the average classification accuracies of K-NN and Naive Bayes for FRGNN3 are superior to all algorithms. The entropy values for FRGNN3 are seen to be lower than all the algorithms. As the FRGNN3 is obtained by integration of fuzzy rough set and neural networks, the FRGNN3 has taken the highest CPU time for Wisconsin and waveform and the second highest CPU time for multiple features data (see the last column of Table 4.7).

Table 4.8 Comparison of UFSFS and FRGNN for microarray gene expression datasets

Data set	Method	Entropy	β-index	DB-index	Fuzzy rough entropy	CPU time in sec.
Cell Cycle ($Z = 45$)	UFSFS	0.244304	0.0883736	5.575267	0.155386	0.6905
	FRGNN3(8)	0.243300	0.082622	5.297885	0.145353	60.7200
Yeast Complex ($Z = 40$)	UFSFS	0.244290	0.095343	3.473932	0.138780	0.6136
	FRGNN3(8)	0.243914	0.109309	3.349231	0.126026	30.0960
All Yeast ($Z = 40$)	UFSFS	0.244092	0.061760	17.747542	0.085130	2.3817
	FRGNN3(8)	0.243497	0.068388	10.903942	0.109613	586.8830

Microarray gene expression data: The importance of the selected features of microarray gene expression data is analyzed using the entropy measure. Moreover, the data with the selected features are grouped into clusters using self-organizing map. The clustering solutions are then evaluated using the β-index, DB-index and fuzzy rough entropy.

The results of the FRGNN3, as compared to UFSFS, for Cell Cycle, Yeast Complex and All Yeast microarray gene expression data are provided in Table 4.8. It may be noted that, Relief-F and SFS can be applied on data sets when the data contains class information (each pattern belongs to a single class only). Here, these methods cannot be applied on microarray gene expression data sets as the class information does not exist with the data (each gene belongs to multiple functional categories according to Munich Information for Protein Sequences (MIPS)).

For Cell Cycle data, the values of entropy, DB-index and fuzzy rough entropy are seen to be lower for FRGNN3 than those of UFSFS. The value of β-index is higher UFSFS than FRGNN3. This means, for Cell Cycle data, FRGNN3 is superior to UFSFS in terms of entropy, DB-index and fuzzy rough entropy whereas the converse is true only in terms of β-index.

For All Yeast data, on the other hand, FRGNN3 is seen to perform better in terms of entropy, β-index and DB-index except fuzzy rough entropy.

Interestingly, for Yeast Complex data, FRGNN3 is superior to UFSFS in terms of entropy, β-index, DB-index and fuzzy rough entropy. While the overall performance of the FRGNN3 has an edge over UFSFS, it requires more CPU time as evident from Table 4.8.

Mean Square Error: The mean square error of FRGNN3 for iris data is shown in Fig. 4.15. It can be observed that the error value is 0.09 for the first iteration and is gradually decreased with the increasing number of iterations. Finally, FRGNN3 converges to a minimum value at 2000th iteration for iris data.

Fig. 4.15 Variation of mean square error with number of iterations of FRGNN3 for iris data

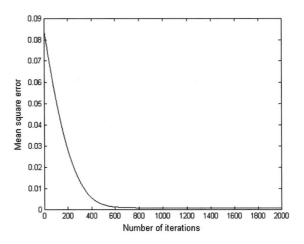

Fig. 4.16 Variation of mean square error with number of iterations of FRGNN3 for waveform data

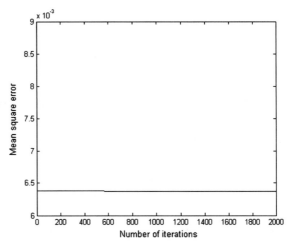

Fig. 4.17 Variation of mean square error with the number of iterations of FRGNN3 for Yeast Complex data

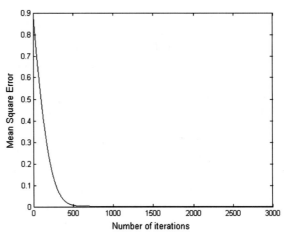

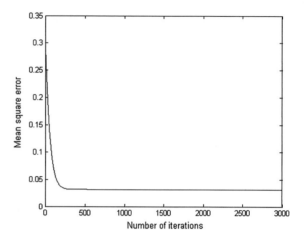

Fig. 4.18 Variation of mean square error with the number of iterations of FRGNN3 for All Yeast data

The mean square error of FRGNN3 for waveform data is shown in Fig. 4.16. The mean square error is initially observed to be 0.00863 at zeroth iteration and is seen to be 0.0064 after few iterations. Thereafter, this value remains the same for FRGNN3 for all the iterations.

For Yeast Complex and All Yeast microarray gene expression data, FRGNN3 converges to the minimum values at 3000 iterations as shown in Figs. 4.17 and 4.18, respectively. The minimum error values signify that FRGNN3 is convergent and its connection weights are approximated. The features corresponding to higher weight values are selected as higher weight value indicates the corresponding feature is better for selection.

4.7 Conclusion

This chapter addressed the problem of feature selection. New notions of lower and upper approximations of a fuzzy rough set are developed by using fuzzy logical operators. The method of designing the FRGNN3 is described by integrating granular computing and neural networks in a soft computing framework. The granulation structures are based on the concept of fuzzy sets. These structures are also used to determine a decision system to extract domain knowledge about data. The domain knowledge, in terms of dependency factors that represent fuzzy rules, is incorporated into a three layered network as its initial connection weights. The fuzzy rules can be represented as granules. The network is trained through minimization of the feature evaluation index in an unsupervised manner. After training is completed, the importance of individual features are obtained from the updated weights between the nodes of the hidden layer and output layer of FRGNN3. A feature corresponding to higher weight value, between hidden and output layers, is selected.

The performance of FRGNN3, in terms of the average percentage of classification accuracies of K-NN and Naive Bayes and the value of entropy, is found to be superior to UFSFS, SFS and Relief-F. These are shown on real life data sets with different dimensions, ranging from 4 to 649. The variation in the mean square error of FRGNN3 for different numbers of iterations is provided for different number of data sets. Superior results of FRGNN3 as compared to UFSFS, in terms of β-index, DB-index and fuzzy rough entropy, are shown for the microarray data with different dimensions, ranging from 79 to 93.

The weakness of the FRGNN3 is that it takes higher computational time than the related methods for feature selection tasks. The selection of α resulting the granulation structures and the number of nodes in the input layer is crucial.

The FRGNN3 can be recommended for feature selection when the data has overlapping classes, with no class information (class labels) and a large number of features. Therefore it has significance in handling large real life data sets. Although, both FRGNN3 and UFSFS are unsupervised feature selection algorithms, the performance of FRGNN3 is superior to UFSFS for all the data sets because the concepts of fuzzy set and fuzzy rough set are used for designing FRGNN3.

The method FRGNN3 ranks the features individually based on their corresponding significance. It could be an interesting study to find the best set, of given cardinality, of features in combination.

References

1. Davies, D.L., Bouldin, D.: A cluster separation measure. IEEE Trans. Pattern Anal. Mach. Intell. **1**(2), 224–227 (1979)
2. Devijver, P.A., Kittler, J.: Pattern Recognition: A Statistical Approach. Prentice Hall, Upper Saddle River (1982)
3. Ganivada, A., Ray, S.S., Pal, S.K.: Fuzzy rough granular self-organizing map and fuzzy rough entropy. Theoret. Comput. Sci. **466**, 37–63 (2012)
4. Ganivada, A., Ray, S.S., Pal, S.K.: Fuzzy rough sets, and a granular neural network for unsupervised feature selection. Neural Netw. **48**, 91–108 (2013)
5. Kira, K., Rendell, L.A.: A practical approach to feature selection. In: Proceedings of 9th International Conference on Machine Learning, vol. 2, no. 6, pp. 249–256 (1992)
6. Mitra, P., Murthy, C.A., Pal, S.K.: Unsupervised feature selection using feature similarity. IEEE Trans. Pattern Anal. Mach. Intell. **24**(4), 301–312 (2002)
7. Newman, D.J., Hettich, S., Blake, C.L., Merz, C.J.: UCI repository of machine learning databases. Department of Information and Computer Science, University of California, Irvine, CA. http://archive.ics.uci.edu/ml/ (1998)
8. Pal, S.K., Dasgupta, B., Mitra, P.: Rough self-organizing map. Appl. Intell. **21**(1), 289–299 (2004)
9. Pal, S.K., De, R.K., Basak, J.: Neuro-fuzzy feature evaluation with theoretical analysis. Neural Netw. **12**, 1429–1455 (1999)
10. Pal, S.K., De, R.K., Basak, J.: Unsupervised feature evaluation: a neuro-fuzzy approach. IEEE Trans. Neural Netw. **11**(2), 366–376 (2000)

11. Pal, S.K., Ghosh, A., Shankar, B.U.: Segmentation of remotely sensed images with fuzzy thresholding, and quantitative evaluation. Int. J. Remote Sens. **21**(11), 2269–2300 (2000)
12. Parthalain, N.M., Jensen, R.: Unsupervised fuzzy-rough set-based dimensionality reduction. Inf. Sci. **229**(3), 106–121 (2013)
13. Verikas, A., Bacauskiene, M.: Feature selection with neural networks. Pattern Recogn. Lett. **23**(11), 1323–1335 (2002)

Chapter 5
Granular Neighborhood Function for Self-organizing Map: Clustering and Gene Selection

5.1 Introduction

In the previous chapters various techniques involving granular neural networks for classification, clustering and feature selection in bioinformatics tasks are described. Although these techniques are very useful in identifying new patterns and improving the accuracy of results, the domain of generating new biological insights as well as new knowledge remained unexplored. The present chapter and subsequent two chapters deal with such issues. While in this chapter and Chap. 6 not only deal with new computational models but also provide new biological knowledge about diseases and function related genes, Chap. 7 provides a survey of existing techniques for RNA (Ribo nucleic acid).

A method of feature selection in real-life data sets, using fuzzy rough granular neural network and minimization of feature evaluation index, is depicted in Chap. 4. The main principle behind this method is based on minimization of error corresponding to a set of selected features obtained at output node, during training. The error corresponding to the selected features reflects the decrease in intra cluster distances and increase in inter cluster distance. Based on this principle, the network ranks the features for selection without considering the actual class labels of data. In this chapter, a feature is represented as a gene in microarray data and methods are developed for clustering samples (patients/patterns) and thereby selection/ranking the genes (features) based on the cluster information. These methods also do not use any class label information of data.

As mentioned in Sect. 1.6, microarray gene expression data are of two types, functional and diseased data. While, the main task in functional data is to predict the biological function/pathway of unclassified genes, in diseased data the main problem is to find the relevant genes, among a large set of genes, associated with a particular disease (say, cancer). The complexity increases further if there are redundant genes in the diseased data. The set of relevant genescan be used to identify unknown

S.K. Pal et al., *Granular Neural Networks, Pattern Recognition and Bioinformatics*, Studies in Computational Intelligence 712, DOI 10.1007/978-3-319-57115-7_5

genes with similar features by using a classifier. Hence, selection of relevant genes is an important task for designing an efficient classifier. Irrelevant genes may lead to poor performance of the classifier. In pattern recognition framework the diseased microarray data can be viewed as a data with larger number of features (genes) and small number of samples (patients/patterns/objects). In this chapter, the problem of finding relevant genes is handled with an unsupervised granular neural network using fuzzy rough sets. In the process a clustering algorithm is developed to extract the domain knowledge about the data. The resultant clusters are utilized in formulation of an algorithm for ranking the genes and their selection.

Different computational methods for gene analysis from microarray data can be found in [27]. Methods like c-means, c-medoids, fuzzy c-means, and self-organizing map (SOM), etc., can be used to group the similar type of samples or genes in microarray data. Methods based on rough fuzzy sets like rough fuzzy possiblistic c-means (RFPCM) [16], rough possibilistic c-means (RPCM) [16], and robust rough fuzzy c-means algorithm (RFCM) [18] for clustering samples are used. Here, the rough fuzzy sets are used to define probabilistic and possibilistic membership values to patterns in set. Models [9, 21] use the concept of information granules for extracting the domain knowledge about data. In the process, [21] uses rough sets whereas [9] uses fuzzy rough sets; thereby making the latter model superior in terms of handling uncertainty. The SOM based models [9, 21] for clustering real life data, including microarray, employ Gaussian function for defining the neighborhood. In designing the granular self-organizing map (GSOM [28]), the other aspects of rough sets, namely, lower and upper approximations, are exploited in defining the neighborhood function of SOM based on the concepts of fuzzy rough set for better modeling of the overlapping regions. The design principle of GSOM is described in this chapter with adequate experimental results in the process of developing a feature selection method.

Gene selection may be viewed as feature selection in pattern recognition framework. The existing methodologies for feature selection involve both classical and modern approaches. A rough set based method for clustering the genes (features) and selecting the genes in best cluster, in terms of a cluster validation measure, is developed in [4]. In [17], information measures are developed using fuzzy rough sets. The method involves maximization of relevance and minimization of redundancy among selected genes. In [22], an unsupervised feature selection network based on minimization of fuzzy feature evaluation index using gradient decent method is developed. Fuzzy rough methodologies for dimensionality reduction are depicted in [14, 26]. While the method in [14] is a supervised one, [26] works in an unsupervised manner. A popular classical approach for feature selection is reported in [19] where the feature having the farthest distance from its neighboring features in a cluster, evolved by K-NN principle, is selected. A method of evolutionary rough gene selection is described in [2] by integrating genetic algorithm and rough set. In the process, the concepts of evolutionary theory like crowding distance measure, selection, crossover and mutation operators are used. A new fitness function to select the relevant genes is defined using attribute redcuts generated using rough sets. The biological relationships among the selected genes, in terms of gene ontology, are found using FatiGO genome database (http://fatigo.bioinfo.cnio.es) [1]. [24] discusses a neighborhood rough set based feature selection method. Its computation involves two factors, namely, distance function and parameter. While the distance function controls the size of a granule, the parameter changes the shape of the granule. By varying the values of these two factors, different sets of features are selected. The set of features which provides superior results in terms of the feature evaluation measures is selected. Besides gene selection, fuzzy rough sets are also used in selection of MicroRNAs (miRNAs), an important indicator of cancers. A fuzzy-rough entropy measure (FREM) is developed in [20] which can rank the miRNAs and thereby identifying the relevant ones. FREM is used to determine the relevance of a miRNA in terms of separability between normal and cancer classes. While computing the FREM for a miRNA, fuzziness takes care of the overlapping between normal and cancer expressions, whereas rough lower approximation determines their class sizes. Here, rough lower approximations in terms of membership values to cancer and normal classes signify the respective class information granules. MiRNAs are sorted according to the highest relevance (i.e., the capability of class separation) and a percentage among them is selected from the top ranked ones. FREM is also used to determine the redundancy between two miRNAs and the redundant ones are removed from the selected set, as per the necessity. Superiority of the FREM as compared to related methods is demonstrated on six miRNA data sets in terms of sensitivity, specificity and F score. A method of gene selection, namely unsupervised fuzzy rough feature selection (UFRFS [28]), based on the concept of fuzzy rough sets and the output clusters of GSOM [28] is described in detail with experimental results in this chapter.

This chapter is organized as follows: Sects. 5.2 and 5.3 explain different state-of-the-art and highly cited clustering and feature selection methods. The granular self-organizing map (GSOM) is elaborated in Sect. 5.4. The method of unsupervised fuzzy rough feature selection (UFRFS) is illustrated in Sect. 5.5. In Sect. 5.6, the comparative results of GSOM in terms four clustering evaluation measures, and UFRFS in terms of classification accuracies and feature evaluation indices, are provided. Section 5.7 concludes the present chapter.

5.2 Methods of Clustering

The granular self-organizing map (GSOM [28]) is compared with seven clustering algorithms like, robust rough fuzzy c-means algorithm (RRFCM) [18], rough fuzzy possibilistic c-means (RFPCM) [16], rough possibilistic c-means (RPCM) [16], fuzzy c-means (FCM), partition around medoids (c-medoids), affinity propagation clustering algorithm (AP method) [7] and SOM. The clustering algorithms used for comparison are either state-of-the-art, or highly cited, or have components similar to the method. While, FCM and SOM fulfill the last two criteria, c-medoids is a highly cited one. The RRFCM, RFPCM and RPCM are state-of-the-art clustering methods and also use similar components, like fuzzy sets and rough sets, in a different manner. AP method is the state-of-the-art algorithm and uses the concept of numerical methods. The performance of all the clustering algorithms is evaluated with four different cluster validation measures, β-index [23], DB-index [5], Dunn-index [6] and fuzzy rough entropy (FRE) [9]. For a cluster, the lower values of DB-index and FRE and higher values of β-index and Dunn-index signify that the cluster is better. The β-index, DB-index, Dunn-index and FRE evaluate the quality of clustering solutions in different ways without using any class label information. While the first three indices are widely studied and evaluate the compactness of output clusters in terms of low intra-cluster distances and high inter cluster distances, the FRE is a recent one [9] and it is a rough-fuzzy set based measure that evaluates the clusters in terms of minimum roughness ($1 - \frac{|\text{lower approxi.}|}{|\text{upper approxi.}|}$). Further, DB-index and Dunn-index are based on maximum or minimum principle computed over distance between clusters and variance, taking two clusters together. On the other hand, β-index considers only the variance computed over individual class and the overall feature space. Let us explain briefly the clustering method RFPCM [16] as follows, before describing the granular self-organizing map (GSOM) [28] in Sect. 5.4.

5.2.1 Rough Fuzzy Possibilistic c-Means

The method of rough fuzzy possibilistic c-means is based on combination of fuzzy c-means algorithm (FCM) and rough possibilistic c-Means (RPCM). In FCM and RPCM, c patterns are selected randomly as the initial means of c-clusters. The probabilistic memberships and possibilistic memberships for all pattern belonging to c-clusters are computed in FCM and RPCM, respectively to define the means of the c clusters. Both the probabilistic memberships and possibilistic memberships are involved in defining the rough fuzzy possibilistic c-means.

Let $\underline{A}(C_i)$ and $\overline{A}(C_i)$ be the lower and upper approximations of cluster C_i, and $\mathbf{B}(C_i) = \{\overline{A}(C_i) - \underline{A}(C_i)\}$ denote the boundary region of cluster C_i. The RFPCM algorithm partitions a set of n objects into c clusters by minimizing the objective function

$$
J_{RFP} = \begin{cases} \{w \times \mathbf{A_1} + \bar{w} \times \mathbf{B_1}, \text{ if } \underline{A}(C_i) \neq \emptyset, \mathbf{B}(C_i) \neq \emptyset, \\ \mathbf{A_1}, \qquad\qquad\quad \text{ if } \underline{A}(C_i) \neq \emptyset, \mathbf{B}(C_i) = \emptyset, \\ \mathbf{B_1}, \qquad\qquad\quad \text{ if } \underline{A}(C_i) = \emptyset, \mathbf{B}(C_i) = \emptyset, \end{cases}
\tag{5.1}
$$

where,

$$
\mathbf{A_1} = \sum_{i=1}^{c} \sum_{x_j \in \underline{A}(C_i)} \{a(\mu_{ij})^{m_1} + b(v_{ij})^{m_2}\} \|x_j - v_i\|^2 + \sum_{i=1}^{c} \eta_i \sum_{x_j \in \underline{A}(C_i)} (1 - v_{ij})^{m_2}.
\tag{5.2}
$$

$$\mathbf{B_1} = \sum_{i=1}^{c} \sum_{x_j \in \mathbf{B}(C_i)} \{a(\mu_{ij})^{m_1} + b(v_{ij})^{m_2}\}\|x_j - v_i\|^2 + \sum_{i=1}^{c} \eta_i \sum_{x_j \in \mathbf{B}(C_i)} (1 - v_{ij})^{m_2}.$$

$$(5.3)$$

The parameters w and $\bar{w}$ correspond to the relative importance of lower and boundary regions. The constants a and b define the relative importance of probabilistic and possibilistic memberships. μ_{ij} and v_{ij} represent probabilistic and possibilistic memberships, respectively, of jth pattern of ith cluster.

Solving Eq. 5.1 with respect to μ_{ij} and v_{ij}, we get

$$\mu_{ij} = (\sum_{k=1}^{c}(\frac{d_{ij}}{d_{kj}})^{\frac{2}{m_1-1}})^{-1}, \text{ where } d_{ij}^2 = \|x_j - v_i\|^2. \qquad (5.4)$$

$$v_{ij} = \frac{1}{1 + E}, \text{ where } E = \{\frac{b\|x_j - v_i\|^2}{\eta_i}\}^{1/m_2-1}. \qquad (5.5)$$

That is, the probabilistic membership μ_{ij} is independent of the constant a, while the constant b has a direct influence on the possibilistic membership v_{ij}. The lower approximation of RFPCM influences the fuzziness of final partition. The weights of the objects in lower approximation of a cluster should be independent of other centroids, and clusters, should not be coupled with their similarity with respect to other centroids. In addition, the objects in lower approximation of a cluster should have a similar influence on the corresponding centroid and cluster. The objects in boundary regions should have a different influence on the centroids and clusters. Therefore, in the RFPCM, the membership values of objects in lower approximation are $\mu_{ij} = v_{ij} = 1$, while those in the boundary region are the same as in the fuzzy c-means and possibilistic c-means. Selection of cluster prototypes of RFPCM and mathematical analysis on the convergence of RFPCM are discussed in [16].

5.3 Methods of Gene Selection

Several gene selection algorithms, unsupervised feature selection using feature similarity (UFSFS) [19], unsupervised fuzzy rough dimensionality reduction (UFRDR) [26] and Algorithm 1 which is designed by replacing GSOM with SOM in UFRFS [28], are used to compare the performance of unsupervised fuzzy rough feature selection (UFRFS [28]). Moreover, the unsupervised UFRFS is compared with the supervised fuzzy rough mutual information based method (FRMIM) [17] and correlation based feature selection algorithm (CFS) [11] to demonstrate its effectiveness as compared to supervised methods. Among the gene selection methods, UFSFS and CFS are popular (highly cited) feature selection methods and UFRDR and FRMIM are state-of-the-art methods based on the concepts of fuzzy rough set. The genes selected by these algorithms are evaluated using three different classifiers like, fuzzy rough granular neural network (FRGNN1) [8], fuzzy multi-layer perceptron (FMLP)

[25], and K-nearest neighbor (K-NN), and a feature evaluation index using entropy measure [21]. Two of the feature selection methods, UFSFS [19] (unsupervised method) and FRMIM [17] (supervised method), are described as follows. After that we describe the method UFRFS [28] in detail in Sect. 5.5.

5.3.1 Unsupervised Feature Selection Using Feature Similarity

Unsupervised feature selection using feature similarity (UFSFS) [19] is a popular and well cited work, based on a feature similarity matrix. The similarity matrix is formed by computing similarities between genes using a distance measure. By using K-NN principle, the features are partitioned into distinct groups based on the similarity matrix. For a group, the most representative gene is selected and the remaining genes are discarded. Here, the representative gene means that the feature has the farthest distance from the remaining features in the group. The same process is repeated for the remaining groups until all the genes are either selected or discarded. A reduced feature subset is then constituted by collecting all the representative features from the groups.

5.3.2 Fuzzy-Rough Mutual Information Based Method

The method based on fuzzy rough mutual information (FRMIM [17]) is discussed as follows. Let $\mathbf{G} = \{G_1, \ldots, G_i, \ldots, G_d\}$ denote the set of genes (features) or fuzzy conditional attributes of given microarray data and $\mathbf{S}$ be the set of selected genes. Let the relevance of gene G_i (fuzzy conditional attribute) with respect to class $\mathbf{D}$ (fuzzy decision attribute) be $f(G_i, \mathbf{D})$ and the redundancy between two genes G_i and G_j (fuzzy condition attributes) be $f(G_i, G_j)$. The total relevance of all selected genes is defined as

$$\mathfrak{I}_{relev} = \sum_{G_i \in S} f(G_i, \mathbf{D}), \qquad (5.6)$$

while the total redundancy among the selected genes is

$$\mathfrak{I}_{redun} = \sum_{G_i, G_j \in S} f(G_i, G_j). \qquad (5.7)$$

The relevance of a gene G_i with respect to class $\mathbf{D}$, $f(G_i, \mathbf{D})$, and the redundancy between two genes G_i and G_j, $f(G_i, G_j)$, are determined by computing the fuzzy rough mutual information between G_i and $\mathbf{D}$, and G_i and G_j, respectively. The objective function is defined as

$$\Im = \Im_{relev} - \Im_{redun} = \sum_i f(G_i, \mathbf{D}) - \sum_{i,j} f(G_i, G_j) \tag{5.8}$$

By minimizing the objective function, the genes are selected using the following greedy algorithm [17]:

1. Initialize $G \leftarrow \{G_1, \ldots, G_i, \ldots, G_d\}$, $S \leftarrow \emptyset$.
2. Generate fuzzy equivalence partition matrix, FEPM, as in [17], for each gene $G_i \in G$.
3. Calculate the relevance $f(G_i, \mathbf{D})$ for each gene $G_i \in G$.
4. Select gene G_i that has the highest relevance $f(G_i, \mathbf{D})$ as the first gene. In effect, $G_i \in S$, and $G = G - G_i$.
5. Generate resultant FEPM M_{G_i,G_j}, between selected gene G_i of S and each of the remaining genes G_j of G.
6. Calculate the redundancy $f(G_i, G_j)$ between selected genes of S and each of the remaining genes of G.
7. From the remaining genes of G, select gene G_j that maximizes

$$f(G_i, \mathbf{D}) - \frac{1}{|S|} \sum_{G_i \in S} f(G_i, G_j) \tag{5.9}$$

As a result of that, $G_j \in S$, and $G = G - G_j$.
8. Repeat the aforementioned steps 5–7 until the desired number of genes is selected.

5.4 Fuzzy Rough Granular Neighborhood for Self-organizing Map

5.4.1 Strategy

During training of GSOM for the first iteration, the adjustment of initial weights of the winning neurons and its neighborhood neurons are dependent on the Gaussian neighborhood function where the method of finding the neighborhood neurons is shown in [15]. After the training of GSOM is completed for the first iteration, the output clusters are presented to a decision table as its decision classes. Based on the decision table, a neighborhood function for updating the connection weights, say modified competitive learning procedure, are defined using the lower and upper approximations of fuzzy rough set. Here, information granules, derived from the lower approximations, are employed in the process of developing the neighborhood function. For the remaining iterations, the weights of GSOM are updated through modified competitive learning procedure.

In this section, the concepts of fuzzy rough set and self-organizing map are used to develop a granular self-organizing map (GSOM). In this regard, the input data is initially normalized within 0 to 1.

5.4.2 Normalization of Data

Let $\{x'_{ij}\}, i = 1, 2, \ldots, s; j = 1, 2, \ldots, n$; denote a set of n-dimensional data where s represents the total number of patterns in data. The data is then normalized between 0 and 1 as

$$x_{ij} = \frac{x'_{ij} - x'_{min}}{x'_{max} - x'_{min}}, \tag{5.10}$$

where x'_{min} and x'_{max} are the minimum and maximum values of x'_{ij}, respectively. The resultant normalized data is then used in the process of developing GSOM and the method of gene selection.

5.4.3 Defining Neighborhood Function and Properties

The granular self-organizing map contains input and output layers. The number of nodes in the input layer is set equivalent to the number of features (genes for microarray data) and in the output layer, it is set equal to be c-number of clusters, chosen by user. Let $\{x\} \in \mathbf{R}^n$ denote a set of n dimensional input patterns. The weights, $w_{kj}, k = 1, 2, \ldots, c; j = 1, 2, \ldots, n$, connecting the links between n nodes in the input layer and c nodes in the output layer, are initialized with the random numbers within 0 to 0.5. Let e denote the number of iterations for training GSOM. For the first iteration, $e = 1$, the GSOM uses the conventional self-organizing map (SOM) (see Sect. 3.2, Chap. 3). The following Steps are involved in its training are as follows:

(S1) Present an n-dimensional input vector $x(e)$ at the nodes in the input layer of GSOM.
(S2) Find the Euclidean distances between the input vector, $x_j(e)$, and weight vector w_{kj} for kth output node using Eq. 3.1.
(S3) Find a winning neuron v among all the output neurons, using Eq. 3.2.
(S4) Update the connection weights of the winning neuron v and the neurons lying within Gaussian neighborhood of that winning neuron (Eq. 3.4), based on Eq. 3.6.

The aforesaid Steps are repeated for all the input patterns used for training GSOM. After completing the training of GSOM for the first iteration, it partitions the data into c output clusters. These are employed in defining a new neighborhood function.

It may be noted that the updated connection weights of GSOM, based on the Gaussian neighborhood, may not be efficient for handling the uncertainty in the overlapping regions between the cluster boundaries as there is no concept of fuzzyness in the formation of Gaussian neighborhood. In order to over come this situation, a new neighborhood function for updating the initial weights of GSOM are defined using the lower and upper approximations of fuzzy rough set, based on the clusters of GSOM at the first iteration.

5.4.4 Formulation of the Map

The aforesaid c-clusters obtained at c nodes are labeled with values, representing the crisp decision classes, and presented to a decision table $S = (U, \mathcal{A} \cup \{d\})$. Here, U is a universe that contains all patterns from c-clusters. The conditional attributes $\mathcal{A} = \{a_1, a_2, \ldots, a_n\}$ are represented with n-features of the patterns (n number of genes in microarray data). The crisp decision classes are denoted by $d_k, k = 1, 2, \ldots, c\}$ where c number of decision classes represent the c-clusters and d_k is kth decision class under the decision attribute $\{d\}$. Let $A \subseteq U$ denote a set, representing a cluster. A fuzzy reflexive relation R_a, between any two patterns x and y in U, with respect to a quantitative attribute $a \in \mathcal{A}$ is defined in Eq. 2.43 (see Sect. 2.9.1, Chap. 2). For an attribute $a \in \{d\}$, the decision classes in fuzzy case are defined in Eq. 3.16 (see Sect. 3.7.4, Chap. 3).

(S1) For every attribute a_j, $j = 1, 2, \ldots, n$, compute fuzzy reflexive relational matrix, using Eq. 2.43.

(S2) Compute membership values for belonging to lower and upper approximations of set A, using Eqs. 4.11 and 4.14, based on the fuzzy reflexive relational matrix corresponding to attribute a_j. These lower and upper approximations are denoted by $L_{kj}(A)$ and $U_{kj}(A)$, $k = 1, 2, \ldots, c$, respectively.

(S3) For every attribute a_j, $j = 1, 2, \ldots, n$, find the averages of all membership values in lower approximation and in upper approximation of the set A, and represent by $L_{kj}^{avg}(A)$ and $U_{kj}^{avg}(A)$, $k = 1, 2, \ldots, c$, respectively.

(S4) For every attribute a_j, $j = 1, 2, \ldots, n$, find boundary region of the set A, denoted by $B_{kj}^{avg}(A)$, by computing $(U_{kj}^{avg}(A) - L_{kj}^{avg}(A))$, $k = 1, 2, \ldots, c$.

(S5) Define the neighborhood function NH_{kj} for the kth node, $k = 1, 2, \ldots, c$, as

$$NH_{kj} = e^{-(L_{kj}^{avg} + B_{kj}^{avg})^2/2}, \ j = 1, 2, \ldots, n. \tag{5.11}$$

Variation of NH_v for different values of $(L_{avg}(A) + B_{avg}(A))$, within 0 to 1, is shown in Fig. 5.1. The significance of the neighborhood function is that the uncertainty arising in overlapping regions between cluster boundaries is handled efficiently.

The sum of all membership values in the lower approximation and boundary region of a neuron v is then computed for defining its neighborhood neurons as

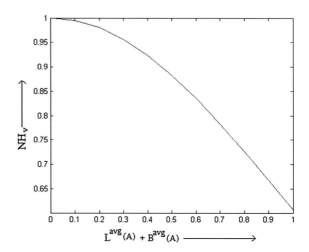

Fig. 5.1 Variation in the neighborhood function NH_v for different values of $(L^{avg}(A) + B^{avg}(A))$, chosen between 0 and 1

$$LBS_k = \sum_{j=1}^{n}(L_{kj}^{avg} + B_{kj}^{avg}). \tag{5.12}$$

Now, a neuron k is considered as the neighborhood neuron of v, when $LBS_k <= LBS_v$ where the LBS values are different for each neuron and dependent on the number of patterns assigned to that cluster.

5.4.5 Algorithm for Training

For the second and subsequent iterations, the following steps are repeated for all input patterns used in training of GSOM.

 (i) Present an n-dimensional input vector $x(e)$ at the nodes in the input layer.
 (ii) Compute the Euclidean distance, using Eq. 3.1, between the input vector $x(e)$ and the weight vector initialized to every node in the output layer.
 (iii) Find a winning neuron v, using Eq. 3.2.
 (iv) Find neurons which are lying within the neighborhood of v, using Eq. 5.12.
 (v) Modify the connection weights of the winning neuron v and its neighborhood neurons using

$$w_{kj}(e+1) = \begin{cases} w_{kj}(e) + \alpha(e)NH_{vj}(t)(x_j(e) - w_{kj}(e)), \\ \qquad\qquad if \quad k \in NH_{vj}(e), \\ w_{kj}(e), else, \end{cases} \tag{5.13}$$

where $\alpha(e)$ is a learning rate and is defined as

$$\alpha(e) = \alpha_0 exp(-\frac{e}{\tau_2}), e = 2, 3 \ldots, \tag{5.14}$$

where τ_2 is another time constant. The value of α_0 is chosen within 0 to 1. The value of $\alpha(e)$ will be monotonically decreased, while the number of iterations, e, is gradually increased. It may be noted that, a neighborhood function is newly defined for every iteration (other than the first iteration), based on the output clusters of GSOM, attained in the previous iteration, using the procedure in Sect. 5.4.4.

Properties: The properties of GSOM are as follows:

Maximality: The neighborhood function attains maximum value 1, when the value of $(L_{kj}^{avg} + B_{kj}^{avg})$ is zero (Fig. 5.1). This implies that, if the membership values in the lower and upper approximations of a set A are zeros then the function attains the maximum value.

Convergence: Several authors examined to establish the existence of convergence for different clustering algorithms, viz., fuzzy c-means, fuzzy curve fitting [30], rough fuzzy possibilistic c-means [16], based on Gauss-Seidel algorithm. The Gauss-Seidel algorithm, representing a set of equations in a matrix form, is assured to converge, when every equation in the matrix is diagonally dominant. Based on this principle, the convergence of the GSOM is discussed. After training of GSOM is completed, a confusion matrix $M_{c \times c}$ of the size $c \times c$ is formed by the output clusters of GSOM where c indicates the number of clusters. Therefore, the matrix M would be

$$M = \begin{pmatrix} A_{11} & A_{12} & \ldots & A_{1c} \\ A_{21} & A_{22} & \ldots & A_{21} \\ A_{c1} & A_{c2} & \ldots & A_{cc} \end{pmatrix},$$

where each row in the matrix M can represent an equation. For l_1 and $l_2 = 1, 2, \ldots, c$, when $l_1 = l_2$ then $A_{l_1 l_2}$ represents the cardinality of the diagonal elements, and when $l_1 \neq l_2$ then $A_{l_1 l_2}$ denote the cardinality of non diagonal elements, in l_1th row of l_1th cluster. When the cardinality of a diagonal element in every row is non zero and the cardinality of non diagonal elements is zero or less than that of diagonal element in that row, then the diagonal element in every row of the matrix M is dominant entry and the algorithm is guaranteed to converge.

5.5 Gene Selection in Microarray Data

Microarray gene expression data comprises, typically, a small number of samples, representing patterns, and a large number of genes, indicating features/attributes, with most of them to be redundant or irrelevant genes. The selection of the relevant genes is very important, when they are used for classification or clustering purpose. In this regard, the GSOM is first used for clustering gene expression data and the resultant clusters are further employed in development of an unsupervised fuzzy rough feature selection (UFRFS) algorithm. Here, the feature is represented with a gene in microarray data. The steps involved in UFRFS are as follows:

S1 Partition all samples of gene expression data into c-clusters/granules using GSOM.
S2 Present the c-granules with labelled values, representing c-crisp decision classes, to a decision table $S = (U, A \cup \{d\})$, $A = \{a_1, a_2, \ldots, a_n\}$.
S3 Find fuzzy decision classes, using Eq. 3.16 (see Sect. 3.7.4, Chap. 3), corresponding to the crisp decision classes, based on the decision table S.
S4 Compute fuzzy reflexive relational matrix, using Eq. 2.43 (see Sect. 2.9.1, Chap. 2), corresponding to attribute a_j, $j = 1, 2, \ldots, n$.
S5 Calculate membership values of the patterns in a set (cluster) for belonging to lower approximation (Eq. 4.11), using the fuzzy reflexive relational matrix and the fuzzy decision classes, corresponding to attribute (gene) a_j, $j = 1, 2, \ldots, n$.
S6 Find a dependency value, using Eq. 2.38, for every conditional attribute (gene) a_j, $j = 1, 2, \ldots, n$, where Eq. 2.37 is used in computing the dependency value.
S7 Arrange the genes in creasing order according to their dependency values.
S8 Select the top-Z number of genes with lower dependency values and, further, use in experiments.

Note that, the membership values of samples in a cluster for belonging to the lower approximation are computed using the membership values of the samples in fuzzy reflexive relation and fuzzy decision classes. When the samples come from its actual cluster then the fuzzy decision values of the samples, lying within 0 to 1, are used as membership values and when the samples come from different cluster (other than the actual cluster) then 1 is assigned as the membership values. Hence, Eq. 4.11 provides lower membership values to the samples belonging to its actual cluster. As the dependency value for every gene is computed by taking the average of all membership values of the samples in the positive regions of the clusters which is dependent on lower approximations a lower dependency value is chosen as the criterion for gene selection.

(a) *Algorithm 1*: A gene selection algorithm, say Algorithm 1, is developed by replacing GSOM with the conventional SOM in UFRFS while the gene selection procedure remains same. The steps involved in the method are described as follows:

(S1) Use self-organizing map (SOM) [15] and partition all samples of gene expression data into c-clusters. Train SOM through competitive learning.
(S2) Use the steps S2 to S8 in UFRFS as mentioned above.

5.6 Experimental Results

The granular self-organizing map (GSOM) and the unsupervised fuzzy rough feature selection (UFRFS) algorithm are implemented in c-programming language and executed using Linux operating system installed in a HP computer, Intel Core i7 CPU

Table 5.1 Characteristics of data

Data	No. of patterns	No. of genes	No. of classes
ALL and AML data [10]	27	5000	2
Prostate cancer [29]	102	12600	2
Lung cancer [3]	197	1000	4
Multiple tissue	103	5565	4
Type (Multi-A) [12]			

880 @ 3.07 GHZ processor and 16 GB RAM. It is configured with Intel Core i7 CPU 880 @ 3.07 GHZ processor and 16 GB RAM. The experiments are carried out on microarray gene expression data with small number of samples (patterns) small number of samples (patients) ranging from 27 to 197 and large number of genes ranging from 1000 to 12600. The characteristics of gene expression data are shown in Table 5.1.

5.6.1 Results of Clustering

Before we explain the results of clustering, the formation of GSOM, involving the newly defined neighborhood function and an equation for updating the connection weights, is explained for AML and ALL data, as an example.

Example:

The number of input nodes in GSOM is considered as 5000 as there are 5000 genes in AML and ALL data. The number of nodes in the output layer is initially set equal to 2. The initial connection weights between nodes in the input and hidden layers are initialized by random numbers chosen within 0 to 0.5. Then the GSOM is trained using normalized data within 0 to 1. For the first iteration ($e = 1$), the GSOM partitions the data into two granules ($c = 2$), based on the competitive learning of self-organizing map (see Steps S1 to S4 in Sect. 5.4.3).

For the second iteration and thereafter, the training of GSOM is performed using the algorithm in Sect. 5.4.5. The samples in the two clusters, obtained from the first iteration, are labelled with integers 1 and 2 indicating two crisp decision classes. The data is then presented to a decision table. The fuzzy decision classes are computed in accordance to the crisp decision classes using Eq. 3.16, Sect. 3.7.4 (see Chap. 3). Based on the decision table, the neighborhood function for every neuron is defined using the procedure in Sect. 5.4.4, thereby an equation (Eq. 5.12) for finding the neighborhood neurons and an equation (Eq. 5.13) for updating the connection weights are defined.

5.6.1.1   Selection of Parameters of GSOM and SOM

Five parameters, e.g., initial learning α_0, iteration t, time constants τ_1 and τ_2 and initial radius σ_0, are used in the training of GSOM and SOM. Here, τ_2 is set equal to the total number of iterations, τ_1 (Eq. 3.5) is dependent on τ_2, and σ_0 (Eq. 3.5) is chosen as the maximum number of neurons in either row or column of the output layer. Considering initial number of clusters $c = 2$, and $\sigma_0 = 2$, we proceed as follows for determining α_0 for different values of e.

For both GSOM and SOM, the value of α_0 is varied within 0 to 1 in steps of 0.05 for a particular number of iterations (e). Then we repeat the same process for other values of e ranging from 10 to 100 in steps of 10. Figure 5.2 shows the variation of clustering measure Dunn-index with α_0 for various values of e. For clarity of presentation, only some curves corresponding to selected α_0 and e are provided. It is observed that the best results for both GSOM and SOM are obtained for $\alpha_0 = 0.3$ and $e = 30$ for AML and ALL data and hence, dotted vertical line is drawn in the figure at the value of $\alpha_0 = 0.3$. The figure also shows the selected numbers of iterations within parenthesis.

For choosing the number of clusters (c), we varied c from 2 to $\sqrt{s}$ in steps of 1 where s is the number of samples in the data. For each value of c, we repeated the aforementioned procedure for different values of α_0 and e, and the top 3 results (for 3 different c values), including the values of α_0 and e for all the data sets, in terms of DB-index, Dunn-index and FRE, are shown in Table 5.2. It is clear that the best results of GSOM and SOM for AML and ALL data are achieved at $c = 2$. Moreover, the GSOM performs better than SOM in terms of DB-index, Dunn-index and FRE. For the remaining data sets, the best c values are marked in bold. Based on the results, it can be concluded that the DB-index, Dunn-index and FRE for all data sets conform c value equal to the number of clusters truly present in the data. Any value other than the true c value makes clustering indices change sharply.

Fig. 5.2 Variation of Dunn-index with the values of learning parameter (α_0) for a fixed no. of iterations using GSOM and SOM for ALL and AML data

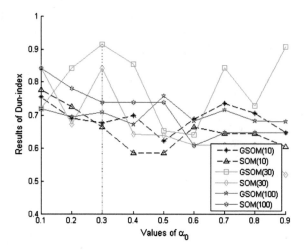

Table 5.2 Comparison of the clustering results of GSOM and SOM. Here, c = Number of clusters, t = Number of iterations, α_0 = Learning parameter

Data set	Method	DB-index	Dunn-index	Fuzzy rough entropy	DB-index	Dunn-index	Fuzzy rough entropy	DB-index	Dunn-index	Fuzzy rough entropy
ALL and AML	GSOM	$c = 2$ 2.0821	0.9150	0.3659	$c = 4$ 2.4707	0.8304	0.4176	$c = 6$ 2.4951	0.5450	0.4370
		$t = 30, \alpha_0 = 0.3$			$t = 10, \alpha_0 = 0.5$			$t = 5, \alpha_0 = 0.8$		
	SOM	2.3074	0.8438	0.3751	2.5626	0.5934	0.4434	2.7112	0.4170	0.4494
		$t = 30, \alpha_0 = 0.3$			$t = 150, \alpha_0 = 0.62$			$t = 200, \alpha_0 = 0.95$		
Prostate cancer	GSOM	1.5127	0.9533	0.2939	2.0575	0.5055	0.3024	2.9182	0.3615	0.2984
		$t = 30, \alpha_0 = 0.6$			$t = 30, \alpha_0 = 0.6$			$t = 15, \alpha_0 = 0.82$		
	SOM	1.6772	0.9190	0.3147	2.1854	0.3473	0.3245	2.9891	0.2814	0.3572
		$t = 100, \alpha_0 = 0.95$			$t = 100, \alpha_0 = 0.95$			$t = 300, \alpha_0 = 0.95$		
Lung cancer	GSOM	$c = 4$ 2.1395	0.6940	0.1441	$c = 6$ 3.5201	0.4160	0.1516	$c = 8$ 4.3477	0.3166	0.1522
		$t = 8, \alpha_0 = 0.05$			$t = 10, \alpha_0 = 0.2$			$t = 10, \alpha_0 = 0.2$		
	SOM	2.2199	0.6829	0.1526	3.9848	0.2860	0.1773	4.4960	0.2726	0.2025
		$t = 300, \alpha_0 = 0.39$			$t = 300, \alpha_0 = 0.9$			$t = 200, \alpha_0 = 0.9$		
Multi-A	GSOM	2.1136	0.7531	0.3348	3.3671	0.4224	0.3468	3.5336	0.3524	0.3477
		$t = 10, \alpha_0 = 0.0686$			$t = 10, \alpha_0 = 0.52$			$t = 10, \alpha_0 = 0.52$		
	SOM	2.1823	0.7071	0.3351	3.1192	0.3951	0.3504	3.9269	0.3520	0.3566
		$t = 200, \alpha_0 = 0.15$			$t = 300, \alpha_0 = 0.9$			$t = 200, \alpha_0 = 0.9$		

5.6.1.2 Visualization of Output Clusters of GSOM and SOM

Figure 5.3 shows the 2D plot of all samples, normalized within 0 to 1, demonstrating the actual categories/groups of AML and ALL data. Figures 5.4 and 5.5 depict the corresponding output clusters ($c = 2$) of GSOM and SOM, respectively.

It is clear from Fig. 5.3 that the boundary region of cluster 1 overlaps with the boundary region of cluster 2. This information is reflected and preserved in Fig. 5.4 for GSOM where the patterns in the overlapping region were exactly clustered as in Fig. 5.3 and the final weights of the winner neurons are represented as '•' and '∗'. In Fig. 5.5, using SOM, 4 overlapping patterns are not exactly the same as in Fig. 5.3. Here 3 patterns actually belonging to cluster 1, indicated with ▷, are assigned to cluster 2 and 1 pattern actually belonging to cluster 2, represented with ◁, is assigned to cluster 1. Moreover, the distance between the winner neurons of GSOM is greater than that of SOM. Here, we want to mention that incorporation of fuzziness in the neighborhood function of GSOM handles the uncertainty arising from overlapping

Fig. 5.3 AML and ALL data in 2D-plane

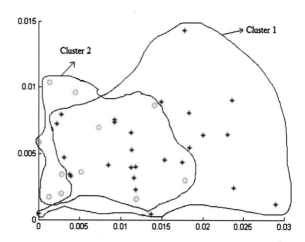

Fig. 5.4 The 2D plot of the output clusters obtained by GSOM for AML and ALL data

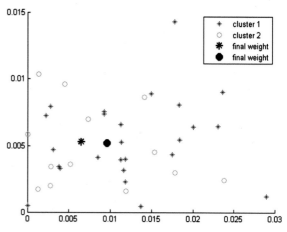

Fig. 5.5 The 2D plot of the output clusters obtained by SOM for AML and ALL data

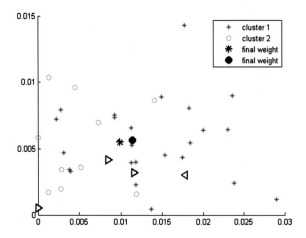

regions, and on the top of this, incorporation of lower and upper approximation of rough set theory further helps in determining the exactness in class shapes; thereby making the GSOM superior to SOM.

5.6.1.3   Comparison of GSOM with Other Clustering Algorithms

The GSOM is compared with some of the sate-of-the-art clustering algorithms like, RRFCM, RFPCM, RPCM, FCM, c-mediods and AP method. The comparative results are shown in Table 5.3. The related parameters of these algorithms are shown in the last column of table. Here, for GSOM and SOM, the values of α_0 and e are provided whereas the values of thresholds for RRFCM (δ and Tr), probabilistic constant and threshold for RFPCM (a and Tr), and threshold value for RPCM (Tr) are shown. The weighting exponents w and $\tilde{w}$, fuzzifiers $\acute{m}_1$ and $\acute{m}_2$, and possiblistic constant b for RRFCM, RFPCM and RPCM are chosen to be 0.5 and 0.5, 2 and 2, 0.7, respectively, for all data sets as mentioned in the related articles [16, 18]. For RRFCM, RFPCM and RPCM, the value of possiblistic constant b is chosen greater than that of probabilistic constant a. The value of c for all the methods is shown within parenthesis in the first column of the table. The maximum value of t for all the state of the art clustering algorithms and all the data sets is set equal to 200. For every data set, the best results in terms of all the clustering indices are shown in bold font in the table.

Table 5.3 shows the results of two microarray data for two output clusters. The results indicate that GSOM performs better in terms of β-index, DB-index, Dunn-index and FRE for all the cases.

The results in Table 5.4 for two data sets with four output clusters show that GSOM outperforms the remaining methods in terms of cluster evaluation measures for all the data except for Multi-A using SOM where β-index is lower. A possible explanation of this may be that β-index does not consider the difference between the means of different clusters, unlike other indices.

Table 5.3 Comparison of clustering algorithms in terms of β-index, DB-index, Dunn-index and FRE for microarray data for $c = 2$. Parameters used are mentioned in 7th column

Data set	Method	β-index	DB-index	Dunn-index	FRE	Parameters
ALL and AML	GSOM	**1.182**	**2.082**	**0.915**	**0.3659**	$30(t), 0.3(\alpha_0)$
	SOM	1.173	2.307	0.845	0.3751	$30(t), 0.3(\alpha_0)$
	RRFCM	1.177	2.204	0.858	0.373	$0.04(\delta), 0.5(Tr)$
	RFPCM	1.173	2.307	0.844	0.3751	$0.6(a), 0.5(Tr)$
	RPCM	1.134	2.725	0.715	0.3801	$0.25(Tr)$
	FCM	1.134	2.731	0.661	0.3886	
	c-medo.	1.087	3.314	0.551	0.3933	
	AP met.	1.089	3.009	0.632	0.3888	
Prostate cancer	GSOM	**1.403**	**1.513**	**0.953**	**0.2939**	$30(t), 0.6(\alpha_0)$
	SOM	1.355	1.677	0.919	0.3147	$100(t), 0.95(\alpha_0)$
	RRFCM	1.377	1.584	0.948	0.2992	$0.1(\delta), 0.05(Tr)$
	RFPCM	1.374	3.024	0.930	0.3014	$0.09(a), 0.54(Tr)$
	RPCM	1.162	3.679	0.737	0.3356	$0.5(Tr)$
	FCM	1.112	5.973	0.566	0.3725	
	c-medo.	1.115	5.612	0.589	0.3384	

Table 5.4 Comparison of clustering algorithms in terms of β-index, DB-index, Dunn-index and FRE for microarray data for $c = 4$. Parameters used are mentioned in 7th column

Data	Method	β	DB	Dunn	FRE	Parameters
Lung cancer	GSOM	**1.627**	**2.139**	**0.694**	**0.1441**	$8(t), 0.05(\alpha_0)$
	SOM	1.593	2.219	0.683	0.153	$300(t), 0.39(\alpha_0)$
	RRFCM	1.571	2.321	0.540	0.1820	$0.21(\delta), 0.005(Tr)$
	RFPCM	1.569	2.456	0.492	0.1881	$0.3(a),0.8(Tr)$
	RPCM	1.566	2.5267	0.438	0.2299	$0.5(Tr)$
	c-medo.	1.534	2.701	0.465	0.2634	
	AP met.	1.358	2.895	0.478	0.2639	
Multi-A	GSOM	1.499	**2.114**	**0.753**	**0.3348**	$10(t), 0.0686(\alpha_0)$
	SOM	**1.500**	2.182	0.707	0.3351	$200(t), 0.15(\alpha_0)$
	RRFCM	1.498	2.328	0.641	0.3372	$0.01(\delta), 0.005(Tr)$
	RFPCM	1.367	2.764	0.645	0.3453	$0.15(a),0.89(Tr)$
	RPCM	1.266	2.992	0.576	0.3499	$0.01(Tr)$
	c-medo.	1.256	2.816	0.560	0.3680	
	AP met.	1.178	2.791	0.499	0.3809	

Table 5.5 Confusion matrix

a) GSOM   b) SOM   c) RRFCM   d) RFPCM

27	0
0	11

24	3
1	10

25	2
1	10

24	3
1	10

e) RPCM   f) FCM   g) c-medoids   h) AP method

22	5
1	10

19	8
2	9

16	11
2	9

25	2
1	10

5.6.1.4   Confusion Matrices for ALL and AML Data Using Different Clustering Methods

The output clusters of GSOM, SOM, RRFCM, RFPCM, RPCM, FCM, c-mediods and AP method, arranged in the form of confusion matrix, are shown in Table 5.5 for AML and ALL data, as an example. In case of GSOM, the samples belonging to output clusters 1 and 2 are seen to be exactly equal to the actual samples, 27 and 11, respectively, in this data; thereby resulting in zero off diagonal elements. This is not true for other methods.

5.6.2   Results of Gene Selection

As mentioned in Sect. 5.5, the UFRFS assigns a dependency value to every gene in a data set. Since the gene with a small dependency value is important for selection, they are arranged in increasing order, according to their dependency values. The top-Z number of genes are then selected based on their classification capability of samples (patients), as measured by classifiers fuzzy granular neural network (FRGNN1), fuzzy MLP (FMLP) and K-NN.

5.6.2.1   Selection of Top Genes

During training of the classifiers, we used 5-fold cross validation, designed with stratified sampling. Training is performed on 4 folds of samples selected randomly. Testing is performed using 1 fold of data. This process is repeated 5 times and the average of the classification accuracies of 5 folds of test data is computed. The average of the classification accuracies, using FRGNN1, FMLP and K-NN, are calculated for top $Z = 10, 20, 30, 40$ and 50 selected genes. The one with the highest accuracy value constitutes the resulting selected genes. As an example, using K-NN, for ALL and AML data, the variation in the average classification accuracies with the numbers of top genes is shown in Fig. 5.6 for all gene selection algorithms. It can be observed from the figure that the highest accuracy for UFRFS is observed at $Z = 20(97.14\%)$. The UFRFS has also achieved the best accuracies for top 20 genes using FMLP (100.00%) and FRGNN1 (100.00%). For the remaining data sets, the values of Z thus obtained are shown in third column of Tables 5.6 and 5.7.

Fig. 5.6 Variation in the classification accuracies obtained using K-NN for top-Z numbers of genes selected by all the algorithms for ALL and AML data

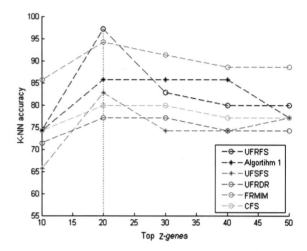

Table 5.6 Comparison of UFRFS with unsupervised gene selection methods in terms of classification accuracy (using FRGNN1, FMLP and K-NN) and entropy of selected genes

Data set	Method	No. of selected genes (N)	FRGNN1	FMLP	K-NN	Entropy	CPU time in Sec.
AML and ALL	UFRFS	20	**100.00**	**100.00**	**97.14**	**0.2177**	40.859
	Algorithm 1	20	97.14	94.28	85.71	0.2401	10.110
	UFRDR	20	94.29	91.43	77.14	0.2421	1368.60
	UFSFS	20	94.29	91.43	82.85	0.2407	95.450
Lung cancer	UFRFS	10	**90.53**	**85.26**	**88.42**	**0.2246**	58.783
	Algorithm 1	10	88.95	85.26	88.42	0.2301	15.075
	UFRDR	10	86.84	83.68	86.32	0.2407	2353.800
	UFSFS	10	85.78	82.63	85.78	0.2407	44.11
Prostate cancer	UFRFS	10	**78.00**	**72.00**	**69.00**	0.2280	441.709
	Algorithm 1	10	63.00	62.00	60.00	0.2232	25.718
	UFRDR	10	63.00	61.00	59.00	**0.2179**	195480.000
	UFSFS	10	62.00	61.00	58.00	0.2279	9582.927
Multi-A	UFRFS	50	**92.63**	**92.63**	**90.57**	**0.2426**	139.069
	Algorithm 1	50	91.58	89.47	88.42	0.2431	26.678
	UFRDR	50	90.52	89.47	90.52	0.2437	27469.8
	UFSFS	50	87.37	86.31	83.13	0.2437	1095.205

Table 5.7 Comparison of UFRFS with supervised FRMIM and CFS in terms of classification accuracy (using FRGNN1, FMLP and K-NN) and entropy of selected genes

Data set	Method	No. of selected genes (N)	FRGNN1	FMLP	K-NN	Entropy	CPU time in Sec.
AML and ALL	UFRFS	20	**100.00**	**100.00**	**97.14**	**0.2177**	40.859
	FRMIM	20	100.00	100.00	94.28	0.2177	4.594
	CFS	20	80.00	74.29	80.00	0.2391	0.100
Lung cancer	UFRFS	10	90.53	85.26	88.42	0.2246	58.783
	FRMIM	10	**95.26**	**95.26**	**93.16**	**0.2245**	0.09
	CFS	10	88.94	85.26	88.42	0.2345	0.110
Prostate cancer	UFRFS	10	78.00	75.00	70.00	**0.2280**	441.709
	FRMIM	10	**88.00**	**86.00**	**85.00**	0.2340	4.275
	CFS	10	71.00	66.00	68.00	0.2306	2543.40
Multi-A	UFRFS	50	92.63	92.63	90.57	0.2426	139.069
	FRMIM	50	**98.95**	**97.90**	**93.68**	**0.2420**	5.233
	CFS	50	98.95	97.90	91.58	0.2423	0.230

5.6.2.2 Comparison of UFRFS with Unsupervised Gene Selection Methods

Table 5.6 also illustrates the results of UFRFS and the other unsupervised methods, Algorithm 1, UFSFS and UFRDR. The results in bold font in the table indicate that the UFRFS performs better than all the unsupervised methods for most of the cases. Here, the performance of UFRFS is equal to that of Algortihm 1 only for the lung cancer using FMLP and K-NN. In other words, the results of unsupervised methods (Table 5.6) indicate that, out of 48 comparisons, UFRFS shows superior performance in 45 cases and equal performance in 2 cases.

In terms of CPU time, Algorithm 1 is better than UFRFS, as expected, as it uses SOM instead of GSOM. Again, UFRFS is superior to UFRDR and UFSFS for all data sets except for lung cancer data where only UFSFS takes lower CPU time. Further, UFRDR takes 54.30 h (with Intel Core i7 CPU 880 @ 3.07 GHZ processor and 16 GB RAM) to complete the execution as compared to 0.1226 h (441.709 s) by UFRFS for prostate cancer data. This limitation of the former for high dimensional data corroborates the earlier finding [26].

5.6.2.3   Comparison of UFRFS with Supervised Gene Selection
Methods

Although UFRFS is an unsupervised method, we compared its performance in a
part of the experiment with two supervised methods namely, fuzzy rough mutual
information based method (FRMIM) [17] and correlation based feature selection
algorithm (CFS) [11]. Table 5.7 describes such comparative results. The comparison
of UFRFS with FRMIM (pairwise comparisons) indicates that out of 16 cases, the
performance of UFRFS is better in 2 cases and equal in 3 cases. On the other hand,
out of 16 cases UFRFS performs better than CFS in 9 cases and equal in 2 cases. For
each of the data sets the best results are shown in bold font in the table.

5.6.3   Biological Significance

The significance of biological categories like biological process, molecular function
or cellular component, of genes selected by the UFRFS is determined using FatiGO
genome database (http://fatigo.bioinfo.cnio.es) [1]. The gene ontology (GO) terms
associated with the set of genes are evaluated using p-value and adjusted p-value
(using Bonferroni correction) and the most significant subcategory under each main
category is provided in Table 5.8 for UFRFS using all the data sets except ethanol
where no significant GO term is obtained for any of the methods. The other significant
subcategories are provided in Table 2 of the supplementary file http://avatharamg.
webs.com/GSOM-UFRFS.pdf. For example, out of 50 selected genes in multi-A
data, 46 genes belong to biological process and out of them 10 genes belong to the
most significant subcategory translational elongation which is shown in Table 5.8.
Two other significant subcategories within biological process are translation and
tissue development with 11 and 9 genes, respectively, and shown in Table 2 of the
supplementary file.

   In the 'biological process' category, the characteristics and relationship of the
genes (selected by the UFRFS) in terms of gene profiles are provided for multi-A
data as an example. The profiles of 10 genes (RPS11, RPL19, RPS16, RPL23A,
RPL18A, RPS27, RPS3A, RPL38, RPS15 and RPS19) belonging to 'translational
elongation' and 11 genes (EIF2AK2, HSPB1, RPS16, RPL8, RPL23A, RPL18A,
RPS27, RPS3A, RPL38, RPL10 and RPS18) belonging to 'translation' for multi-
A data are depicted in Figs. 5.7 and 5.8, respectively. The figures also provide the
average profiles of genes shown in blue line. It is evident from the figures that
the profiles are very similar (correlated) for most of the genes within a particular
subcategory. For example, by considering 10 genes corresponding to Fig. 5.7, the
average of Pearson's correlations for all possible gene pairs (90 pairs excluding self
similarity) is 0.77. Similarly, for 11 genes, corresponding to Fig. 5.8, the average of
Pearson's correlations is 0.76. Similar type of correlation is also observed for other
subcategories as shown in Fig. 5.9 as an example. Here, expression profiles of 9 genes

Fig. 5.7 Plot of the
expression profiles for 10
genes (RPS11, RPL19,
RPS16, RPL23A, RPL18A,
RPS27, RPS3A, RPL38,
RPS15 and RPS19)
belonging to GO biological
process 'translational
elongation' for multi-A data
with 103 samples. The
average of Pearson's
correlations for all possible
gene pairs is 0.77. The
average of expression
profiles is shown in *blue
color*

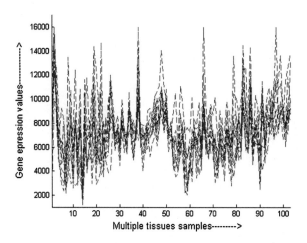

Fig. 5.8 Plot of the
expression profiles for 11
genes (EIF2AK2, HSPB1,
RPS16, RPL8, RPL23A,
RPL18A, RPS27, RPS3A,
RPL38, RPL10 and RPS18)
belonging to GO biological
process 'translation' for
multi-A data with 103
samples. The average of
Pearson's correlations for all
possible gene pairs is 0.76.
The average of expression
profiles is shown in *blue
color*

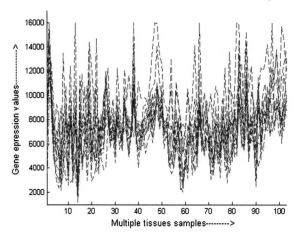

are presented and the average of Pearson's correlations for all possible gene pairs is 0.78. Moreover, all of the gene expressions profiles are seen to be correlated with the average of those profiles.

Further, from the Figs. 5.7 and 5.8 it can be observed that there is also some correlation between the genes in two different subcategories ('translational elongation' and 'translation') within a main category (biological process) and this correlation is less than those of the genes within individual subcategory. This visual interpretation is also supported by the fact when we computed the average of Pearson's correlations as 0.71 (which is less than both 0.77 and 0.76) by considering all the genes in subcategories 'translational elongation' and 'translation'.

Note that, even for two different categories there exist correlation in expression profiles for the genes selected within the same data. It can be observed from Fig. 5.9, which shows expression profile of 9 genes for subcategory 'structural constituent of

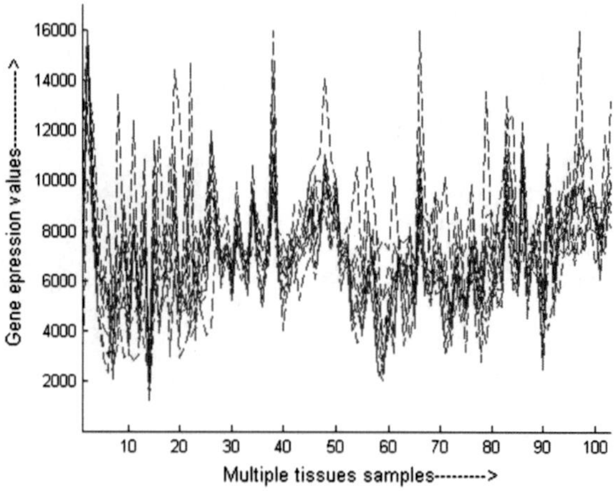

Fig. 5.9 Plot of the expression profiles for 9 genes (RPL8, RPL10, RPS16, RPL23A, RPL18A, RPL38, RPS15, RPS18 and RPS19) belonging to GO molecular function 'structural constituent of ribosome' for multi-A data with 103 samples. The average of Pearson's correlations for all possible gene pairs is 0.78. The average of expression profiles is shown in *blue color*

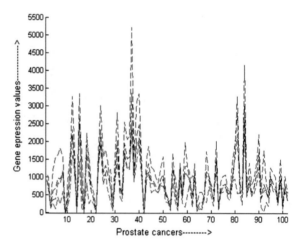

Fig. 5.10 Plot of the expression profiles for 3 genes (37407_s_at, 32755_at and 767_at) belonging to GO cellular component 'contractile fiber' for prostate cancer data with 102 samples. The average of Pearson's correlations for three pairs of genes is 0.88. The average of expression profiles is also shown

ribosome' within main category molecular function, and Fig. 5.7 (described above) that there exists some correlation between the gene profiles for different subcategories using multi-A data. This is also established by the fact when we considered all the genes within subcategory 'structural constituent of ribosome' and 'translational elongation' together and found the average of Pearson's correlations for all possible gene pairs as 0.72.

In another example for prostate cancer data, out of 10 selected genes 3 genes belong to category cellular component 'contractile fiber' (shown in Table 5.8) and 3 genes belong to category cellular component 'actin cytoskeleton'. The profiles of 3 genes belonging to contractile fiber are depicted in Fig. 5.10 using 102 samples. From the figure it can be observed that the profiles of the genes are similar (correlated). Here, the average of Pearson's correlations between three pairs of genes (excluding self similarity) is 0.88.

Table 5.8 Biological significance of genes, selected by the UFRFS

Data set	GO biological category	Gene ontology term	p-value	Adjusted p-value
Multi-A	Biological process	Translational elongation	3.767e-21	8.27e-18
	Biological process	Translational elongation	2.36e-19	5.181e-16
	Molecular function	Structural constituent of ribosome	7.301e-14	7.191e-11
	Cellular component	Cytosolic ribosome	1.796e-25	4.167e-23
Lung cancer	Biological process	Response to wounding	4.189e-17	4.61e-14
	Molecular function	Structural constituent of cytoskeleton	1.022e-12	5.051e-10
	Cellular component	Vesicle	6.167e-18	1.431e-15
Prostate data	Biological process	Formation involved in morphogenesis	1.4480e-005	0.01589
	Cellular component	Contractile fiber	1.6890e-005	0.003917

The functionally enriched biological categories should be related to the cancer type of the data set. In this regard we conducted a literature study. For example, the breast cancer data shows functional enrichment in the categories biological process: 'cellular homeostasis' and cellular component: 'extracellular space'. These functional categories are related to breast cancer according to [13]. In a similar way, the identified functional categories for other data sets are also involved with the related cancer and these are mentioned in Table 3 of the supplementary file http://avatharamg.webs.com/GSOM-UFRFS.pdf.

Table 5.9 provides a biological comparison for the UFRFS, Algorithm 1 and FRMIM using ALL and AML data. The genes selected by UFSFS, UFRDR and CFS are not biologically significant for this data and hence not provided in the table. The p-values and adjusted p-values for GO terms using UFRFS are lower than those of Algorithm 1 and FRMIM for all the biological categories. No significant GO terms belonging to molecular function and cellular component are found using FRMIM.

Table 5.9 Biological comparison of genes selected by UFRFS, Algorithm 1 and FRMIM for ALL and AML data

Data set	Method	GO biological category	Gene ontology term	p-value	Adjusted p-value
ALL and AML	UFRFS	Biological process	Oxygen transport	3.725e-10	8.176e-7
			Gas transport	1.438e-9	1.579e-6
		Molecular function	Oxygen transporter activity	1.23e-10	1.211e-7
			Oxygen binding	2.711e-8	1.335e-5
		Cellular component	Cytosolic part	3.165e-7	7.343e-5
			Extracellular space	3.105e-4	0.03602
	Algorithm 1	Biological process	Oxygen transport	1.996e-7	3.862e-4
			Gas transport	4.988e-7	3.862e-4
		Molecular function	Oxygen transporter activity	9.666e-8	9.521e-5
			Oxygen binding	3.961e-6	1.951e-3
		Cellular component	Cytosolic part	1.107e-5	0.002567
	FRMIM	Biological process	Glial cell differentiation	2.401e-5	0.04159
			Gliogenesis	3.79e-5	0.04159

5.7   Conclusion

This chapter describes a granular self-organizing map (GSOM), designed by integrating fuzzy rough sets and SOM in the soft computing framework, to capture the uncertainty and underlying clusters in the data. Fuzzy rough sets use fuzzy reflexive relation and fuzzy decision classes which deal with the vagueness in the class information. The new neighborhood function in GSOM updates the connection weights associated with the output nodes. Based on the output clusters of GSOM, the dependency factor of each attribute (gene), with respect to all the clusters, is computed using fuzzy rough sets. Lower the dependency degree of a gene is, the higher its relevance is in diagnosing cancer. Using this criterion, genes are ranked for selection.

The superiority of GSOM, as compared to robust rough fuzzy c-means (RRFCM), self-organizing map (SOM), rough fuzzy possiblistic c-means (RFPCM), rough possibilistic c-means (RPCM), fuzzy cmeans (FCM), c-medoids and affinity propagation clustering algorithm (AP method) is demonstrated in clustering microarray gene expression data sets in terms of β-index, DB-index, Dunn-index and fuzzy rough entropy (FRE). Intuitively, for handling data with overlapping patterns (like microarrays), any rough fuzzy clustering (RFC) method, like GSOM, will perform better than SOM, c-medoids, AP method, fuzzy clustering (e.g., fuzzy c-means) and rough clustering method (e.g., RPCM) as RFC has one/two more component/s. Further, conventional SOM has advantages over c-means when the shape of the real clusters in the data is non-circular. Hence, the performance of GSOM is expected to be better than RPCM, FRPCM and RRFCM which uses c-means instead of SOM. Moreover, the granulation structures in GSOM preserve the relative information among patterns by using fuzzy similarity matrix and then transfer it to the neighborhood function through rough approximation operators which is missing in other rough fuzzy clustering methods.

The genes selected by the unsupervised method UFRFS are found to be more relevant than those of the other unsupervised methods, in terms of feature evaluation index and classification accuracy in most of the cases. The UFRFS provides better results than related unsupervised methods and similar results to supervised method CFS, out of two (FRMIM and CFS). The genes selected by the UFRFS are more biologically meaningful for all data sets. ♣

So far in Chaps. 2–5, different granular neural networks for classification, clustering and feature selection (gene selection in microarray) are formulated by the judicious integration of fuzzy sets, fuzzy rough sets with artificial neural networks. Several real life data have been used to demonstrate the effectiveness of these networks. Microarray gene expression data is one of them which is used in Chap. 2 for gene clustering. The following Chaps. 6 and 7 deal entirely with the applications in bioinformatics addressing the problems of gene analysis and RNA (ribo-nucleic acid) secondary structure prediction. Furthermore, the relevance of soft computing (i.e., fuzzy sets, neural networks and genetic algorithms) in predicting RNA structure is explored in the last chapter.

References

1. Al-Shahrour, F., Diaz-Uriarte, R., Dopazo, J.: Fatigo: a web tool for finding significant associations of gene ontology terms with groups of genes. Bioinformatics **20**(4), 578–580 (2004)
2. Banerjee, M., Mitra, S., Banka, H.: Evolutionary rough feature selection in gene expression data. IEEE Trans. Syst. Man Cybern. Part C **37**(4), 622–632 (2007)
3. Bhattacharjee, A., Richards, W.G., Staunton, J., Li, C., Monti, S., Vasa, P., Ladd, C., Beheshti, J., Bueno, R., Gillette, M., Loda, M., Weber, G., Mark, E.J., Lander, E.S., Wong, W., Johnson, B.E., Golub, T.R., Sugarbaker, D.J., Meyerson, M.: Classification of human lung carcinomas by mrna expression profiling reveals distinct adenocarcinomas sub-classes. Proc. Natl. Aca. Sci., USA **98**(24), 13,790–13,795 (2001)
4. Chiang, J.H., Ho, S.H.: A combination of rough-based feature selection and rbf neural network for classification using gene expression data. IEEE Trans. Nanobiosci. **7**(1), 91–99 (2008)
5. Davies, D.L., Bouldin, D.: A cluster separation measure. IEEE Trans. Pattern Anal. Mach. Intell. **1**(2), 224–227 (1979)
6. Dunn, J.C.: A fuzzy relative of the isodata process and its sse in detecting compact well-separated clusters. J. Cybern. **3**(3), 32–57 (1973)

7. Frey, B.J., Dueck, D.: Clustering by passing messages between data points. Science **315**(2), 972–976 (2007)
8. Ganivada, A., Dutta, S., Pal, S.K.: Fuzzy rough granular neural networks, fuzzy granules, and classification. Theoret. Comput. Sci. **412**, 5834–5853 (2011)
9. Ganivada, A., Ray, S.S., Pal, S.K.: Fuzzy rough granular self-organizing map and fuzzy rough entropy. Theoret. Comput. Sci. **466**, 37–63 (2012)
10. Golub, T.R., Slonim, D.K., Tamayo, P., Huard, C., Gaasenbeek, M., Mesirov, J.P., Coller, H., Loh, M.L., Downing, J.R., Caligiuri, M.A., Bloomfield, C.D., Lander, E.S.: Molecular classification of cancer: class discovery and class prediction by gene expression monitoring. Science **286**(15), 531–537 (1999)
11. Hall, M.A.: Correlation-based feature selection for discrete and numeric class machine learning. In: Proceedings of 7th International Conference on Machine Learning, pp. 359–366 (2000)
12. Hoshida, Y., Brunet, J.P., Tamayo, P., Golub, T.R., Mesirov, J.P.: Subclass mapping: identifying common subtypes in independent disease data sets. Blood **11**(2), 1–8 (2007)
13. Ifergan, I., Shafran, A., Jansen, G., Hooijberg, J.H., Scheffer, G.L., Assaraf, Y.G.: Folate deprivation results in the loss of breast cancer resistance protein (bcrp/abcg2) expression: a role for bcrp in cellular folate homeostasis. J. Biol. Chem. **279**(24), 25527–25534 (2004)
14. Jensen, R., Shen, Q.: New approaches to fuzzy-rough feature selection. IEEE Trans. Fuzzy Syst. **17**(4), 824–838 (2009)
15. Kohonen, T.: Self-organizing maps. Proc. IEEE **78**(9), 1464–1480 (1990)
16. Maji, P., Pal, S.K.: Rough set based generalized fuzzy c-means algorithm and quantitative indices. IEEE Trans. Syst. Man Cyern. Part B **37**(6), 1529–1540 (2007)
17. Maji, P., Pal, S.K.: Fuzzy-rough sets for information measures and selection of relevant genes from microarray data. IEEE Trans. Syst. Man Cyern. Part B **40**(3), 741–752 (2010)
18. Maji, P., Paul, S.: Rough-fuzzy clustering for grouping functionally similar genes from microarray data. IEEE/ACM Trans. Comput. Biol. Bioinf. **10**(2), 286–299 (2013)
19. Mitra, P., Murthy, C.A., Pal, S.K.: Unsupervised feature selection using feature similarity. IEEE Trans. Pattern Anal. Mach. Intell. **24**(4), 301–312 (2002)
20. Pal, J.K., Ray, S.S., Cho, S.B., Pal, S.K.: Fuzzy-rough entropy measure and histogram based patient selection for mirna ranking in cancer. IEEE/ACM Trans. Comput. Biol. Bioinf. (2016, accepted). doi:10.1109/TCBB.2016.2623605
21. Pal, S.K., Dasgupta, B., Mitra, P.: Rough self-organizing map. Appl. Intell. **21**(1), 289–299 (2004)
22. Pal, S.K., De, R.K., Basak, J.: Unsupervised feature evaluation: a neuro-fuzzy approach. IEEE Trans. Neural Netw. **11**(2), 366–376 (2000)
23. Pal, S.K., Ghosh, A., Shankar, B.U.: Segmentation of remotely sensed images with fuzzy thresholding, and quantitative evaluation. Int. J. Remote Sens. **21**(11), 2269–2300 (2000)
24. Pal, S.K., Meher, S.K., Dutta, S.: Class-dependent rough-fuzzy granular space, dispersion index and classification. Pattern Recogn. **45**(7), 2690–2707 (2012)
25. Pal, S.K., Mitra, P.: Rough fuzzy MLP: modular evolution, rule generation and evaluation. IEEE Trans. Knowl. Data Eng. **15**(1), 14–25 (2003)
26. Parthalain, N.M., Jensen, R.: Unsupervised fuzzy-rough set-based dimensionality reduction. Inf. Sci. **229**(3), 106–121 (2013)
27. Quackenbush, J.: Computational analysis of microarray data. Nat. Rev. Genet. **2**(6), 418–427 (2001)
28. Ray, S.S., Ganivada, A., Pal, S.K.: A granular self-organizing map for clustering and gene selection in microarray data. IEEE Trans. Neural Netw. Learn. Syst. **27**(9), 1890–1906 (2016)
29. Singh, D., Febbo, P.G., Jackson, K.R.D.G., Manola, J., Ladd, C., Tamayo, P., Renshaw, A.A., D′Amico, A.V., Richie, J.P., Lander, E.S., Loda, M., Kantoff, P.W., Golub, T.R., Sellers, W.R.: Gene expression correlates of clinical prostate cancer behaviour. Cancer Cell **1**, 203–209 (2002)
30. Yan, H.: Convergence condition and efficient implementation of the fuzzy curve-tracing (fct) algorithm. IEEE Trans. Syst. Man Cyern. Part B **34**(1), 210–221 (2004)

Chapter 6
Gene Function Analysis

6.1 Introduction

In Chap. 5, a granular self-organizing network for clustering and a methodology for gene ranking using the concept of granularity are described. Their performance is evaluated using several microarray gene expression data sets where the number of genes is higher than the number of patients/samples. As mentioned there, the gene ranking methodology is useful in identifying new disease related genes and thereby generating new biological insights. However, to identify or predict the function of genes we need datasets dealing with the cellular mechanism of a normal cell rather than diseased condition. The present chapter deals with this issue where functions of unclassified genes are predicted by applying clustering and genetic algorithm based optimal ordering technique [40] on microarray dataset [14, 50] as well as on integrated information from other biological data sources like phenotypic profiles [9], protein sequences [30], Kyoto Encyclopedia of Genes and Genomes (KEGG) pathway information [23] and protein-protein interaction data [42, 43]. Even in a model organism like Yeast, there are more than 1000 genes with unknown biological function defined in Munich Information Center for Protein Sequences (MIPS) [33] and Saccharomyces Genome Database (SGD) [13]. One of the common approaches in predicting function or pathway of unclassified gene involves identifying the group of its closest classified genes and assigning the common biological function or pathway of the group to the unclassified gene, using aforementioned different sources of information. It may be noted that, till now granular computing is not utilized neither in any method described in this chapter nor in any techniques related to gene function prediction, developed so far. This chapter mainly addresses the problem of gene expression analysis using gene ordering, clustering and combinatorial optimization through genetic algorithm. There is also scope to introduce granular computing in the methodologies developed for gene analysis and their function prediction since clustering and prediction tasks are involved in the process.

© Springer International Publishing AG 2017
S.K. Pal et al., *Granular Neural Networks, Pattern Recognition and Bioinformatics*, Studies in Computational Intelligence 712,
DOI 10.1007/978-3-319-57115-7_6

A key step in the analysis of gene function is the identification of groups of genes that manifest similar patterns in terms of their expression or related protein sequence or pathway profile or using all of these information together. This translates to the task of clustering genes using any particular data source or integrated information from multiple data sources. A measure of similarity is defined between pairs of gene which is used for clustering. Clustering methods can be broadly divided into two categories, namely hierarchical and partitive. Hierarchical clustering (e.g., single, complete and average linkage) [5, 8, 14, 24] groups gene expressions into trees of clusters. However, it has a number of shortcomings. It has been noted by statisticians that, hierarchical clustering suffers from lack of robustness, nonuniqueness, and inversion problems that complicate interpretation of the hierarchy [50]. The deterministic nature of hierarchical clustering can cause points to be grouped based on local decisions, with no opportunity to reevaluate the clustering. The resulting trees can lock in accidental features, reflecting idiosyncrasies of the agglomeration rule [50]. Partitive clustering algorithms, such as K-means [19], self-organizing map (SOM) [50], CAST [7], and CLICK [44], separate genes into groups according to the degree of distance or similarity among genes. The shortcoming of partitive clustering is that, relationships among the genes in a particular cluster are lost. Integrating gene ordering with partitive clustering is an approach that is likely to overcome this problem. To obtain the relationship among genes in clusters, a hybrid method [39] is described in this chapter where the gene ordering algorithm FRAG_GALK [40], is used to order genes in clusters obtained from partitive clustering solutions of CLICK [44], k-means [19] and self organizing map (SOM) [50]. For the purpose of comparison, an existing optimal leaf ordering method in hierarchical clustering framework [5] is also used to order genes. The utility of the new hybrid algorithm [39] is shown in improving the quality of the clusters provided by any partitive clustering algorithm by,

- identification of subclusters within big clusters, and
- grouping functionally correlated genes within clusters using MIPS categorization.

Gene expressions provide a relatively new source for gene function prediction, but they alone often lack the degree of specificity needed for accurate prediction. This lack of specificity is due to the drawbacks of microarray technology, and it can be overcome through the incorporation of heterogeneous functional data in an integrated analysis [51]. The working hypothesis for various data source integration is that each set of functional genomics data has an intrinsic error rate and a limited coverage but informs us to some extent about the tendency for genes to operate in the same cellular functions and pathways in the cell. Therefore a more accurate and extensive functional predictions can be achieved by integrating the information from multiple functional genomics data sets, and in this manner the overall functional coupling between genes across a broad set of experiments can be estimated [26]. The present chapter also deals with such an issue where the relevance of each data source is first identified in the common framework of positive predictive value (*PPV*), using yeast GO-Slim process annotations [13], and then integrated to form 'Biological Scores' of genes through a linear combination of weights [41]. The weights are systematically determined by using the same annotations. Finally, the 'Biological Scores' (*BS*) are clustered and functions of unclassified genes are predicted from those clusters which show functional enrichment for a particular function according to Munich Information Center for Protein Sequences (MIPS) [33] database. The superiority of the *BS* as compared to individual datasources and another related method using Bayesian Score [26] is also established in terms of *PPV* of top gene pairs. The effectiveness of the *BS* is shown in predicting the functions of 12 unclassified yeast genes with 0.98 positive predictive value. Those functions are recognized as computational novel predictions in Saccharomyces Genome Database (SGD) [13].

The chapter is organized as follows: Analysis of gene expression, based on normalization, distance measures, gene clustering and ordering and the utility in integrating other relevant data sources with gene expression, are illustrated in Sect. 6.2. Different data sources and the respective pair-wise gene similarity extraction techniques are described in Sect. 6.3. Evaluation of dependence among data sources and relevance of data sources are also discussed. Section 6.4 provides methods of single and multiple data sources for predicting gene function and corresponding results with biological interpretation. Relevance of soft computing and granular networks with gene ordering and clustering is discussed in Sect. 6.5. Conclusions are provided in Sect. 6.6.

6.2 Gene Expression Analysis: Tasks

Various methods are developed for detecting and quantifying the amount of mRNA or gene expression level. These methods take advantage of the sequence complimentary of DNA. The key observation was that single stranded DNA binds strongly to nitrocellulose membranes which prevents strands from disassociating with each other but permits the hybridization to complementary RNA [16]. This led to the development of blotting methods, the first of which combined filter hybridization with gel separation of restriction digests [46]. The blotting methods are serial in nature and the mRNA is measured one at a time. On the other hand, DNA microarrays allow one to interrogate the mRNA population expressed by thousands of genes at once rather than serially as in the blotting methods. Another distinction between DNA microarrays and blotting methods is that the use of impermeable rigid substance such as glass to bind the DNA sequences in microarrays is practically advantageous over porous membranes and gel pads in blotting methods.

The two basic types of commonly used DNA microarrays are spotted arrays and oligonucleotide arrays. In the spotted array methods [14] a large number of cDNAs are prepared from a cDNA library and then spotted onto a glass slide by a robot. Each cDNA corresponds to one gene or one exon in the genome of length 100–1000 bp and also is referred as a probe. Each probe corresponds to a particular spot in a microarray slide. In the course of hybridization, RNA from experimental samples (taken at selected times during the process) is labeled during reverse transcription with the red-fluorescent dye Cy5 and is mixed with a reference sample (cDNA) labeled in parallel with the green-fluorescent dye Cy3 [14]. After hybridization and appropriate washing steps, the arrays are scanned to produce images and the images are further processed by an image analysis program to produce measured red and green foreground and background intensities for each spot on each array. Before the gene expression profiles of the RNA samples can be analyzed and interpreted, the red and green intensities must be normalized relative to one another so that the red/green ratios for all target elements are as far as possible an unbiased representation of relative expression. If R (red) and G (green) are the spot-specific, quantified, fluorescent intensities of the target and reference expression signals respectively, relative gene expression is defined as the log ratio $M = \log_2 \frac{R}{G}$. For microarray data table each cell represents the measured Cy5/Cy3 fluorescence ratio (M value or $\frac{R}{G}$ value) at the corresponding target element [14] obtained from the gene under that experimental condition. All $\frac{R}{G}$ values are log transformed (base 2 for simplicity) to treat inductions or repressions of identical magnitude but with opposite sign.

6.2.1   Preprocessing

Estimating missing values is a classical problem in statistics, and iterative algorithms based on the expectation maximization (EM) algorithm [22] are widely used. Microarray experiments generate data sets with information on the expression levels of thousands of genes in a set of biological samples. Unfortunately, such experiments often produce multiple missing expression values, normally due to various experimental problems. Typical problems include spotting problems, scratches on the slide, dust or hybridization failures. This in turn results in values missing from the gene expression matrix. Some methods for estimation of missing values in microarray data sets are presented in [18]. The methods, based on the least squares principle, are available in a software package called LSimpute and utilize correlations between both genes and arrays. In LSimpute software, two basic methods based on least squares principle, one utilizing correlations between genes (LSimpute gene) and the other utilizing correlations between arrays (LSimpute array), are used to estimate missing values. A robust method for using weighted averages of the estimates from LSimpute gene and LSimpute array into adaptive estimates is also available in LSimpute. Others methods based on nearest neighbor and expectation maximization are available in KNNimpute [51] and EMimpute [18], respectively.

To identify relationship among genes, involved in multiple biological functions or processes, many microarray experiments with different biological origins are conducted. These experiments with multiple microarray slides are sources of nonbiological variation between slides such as dye biases, sample preparation or hybridization differences, scanner calibrations, slide printing variations, volume of initial RNA, etc. Some of these variabilities can be corrected by data normalization before analysis of the data. The main assumption behind normalization of microarray data is that most of the genes on the slide do not change their expression levels and the numbers of up- and down-regulated genes on the array are roughly equal. Most methods try to adjust expression levels of the genes such that the overall average expression remains the same across different arrays. Additional steps in normalization can be performed by removing saturated signals from microarray, background correction, low expression genes correction, etc. In cDNA microarray related investigations, many different methods are developed in order to compensate for dye-effects and other non-biological variations between arrays. The existing normalization methods for microarray data are broadly classified in the following groups [57]:

1. Within-slide normalization: Normalization in specific location

   - Global normalization
   - Intensity dependent normalization
   - Within-print-tip-group normalization

2. Within-slide normalization: Normalization with slide dependent scaling factor

   - Median absolute deviation (MAD)
   - Variance regularization

3. Paired-slides normalization (dye-swap)
4. Between-slide or Multiple-slide normalization

   - Median absolute deviation (MAD)
   - Variance regularization

Global normalization methods assume that the red and green intensities are related by a constant factor k (that is, $R = kG$). Generally the center of the distribution of log-ratios is shifted to zero by subtracting the median or mean of the log-ratios for a particular gene. Global normalization approaches also include rank invariant methods but do not take into account the spatial or intensity dependent die biases. On the other hand, LOWESS (locally weighted scatter plot smooth), a locally weighted linear regression method [10] accounts for such effects and has been proven to be a robust, powerful normalization method for correcting intensity-dependent ratio bias in different types of two color (R and G) microarray experiments [57]. LOWESS is used for microarray data with separate R and G value for each experimental condition. For microarray data sets with given ratio values ($\log_2 \frac{R}{G}$), it is assumed that intensity-dependent ratio bias is corrected by the data providers. Within print-tip-group normalization accounts for differences in the length, opening, and deformation of the tips, used for printing grids in a microarray. After this normalization all the normalized log-ratios from the different print-tip-groups will be centered around zero. However, there is a possibility that the log-ratios from the various print-tip-groups have different spread and there are substantial scale differences between them, because of changes in the photomultiplier tube settings of the scanner or for other reasons. In these circumstances within-slide scale normalization is required. Scale-normalization is a simple scaling of the log-ratios from a series of arrays and the scaling factor is estimated using median absolute deviation (MAD) or variance of log-ratios for individual print-tip-groups. Similar type of scale normalization, using

MAD and variance, is performed in between-slide or multiple-slide normalization when, log-ratios from the various slides (experiments) have different spread.

Paired-slides normalization applies to dye-swap experiments where, two hybridizations are performed for two mRNA samples and the dye assignment is reversed in the second hybridization [57]. A good evaluation of new normalization methods and their comparisons are available in [55, 57], respectively.

A different approach is required to normalize the affymetrix arrays, that use only one channel to measure abundance of mRNA levels. Several methods are developed in this regard. The default method, for minimizing biological and technical variations in GeneChip expression microarray, uses a constant scaling factor (SF), for every gene on an array. The SF is obtained from a trimmed average signal of the array after excluding the 2% of the probe sets with the highest and the lowest values.

6.2.2 Distance Measures

The most popular and probably most simple measures for finding global distance (or similarity) between genes, using gene expression, are the Euclidean distance and Pearson correlation, a statistical measure of linear dependence between random variables.

Let $X = x_1, x_2, \ldots, x_k$ and $Y = y_1, y_2, \ldots, y_k$ be the expression levels of the two genes in terms of log-transformed microarray gene expression data obtained over a series of k experiments. While the Euclidean distance between gene X and gene Y is defined as

$$E_{X,Y} = \sqrt{\{x_1 - y_1\}^2 + \{x_2 - y_2\}^2 + \cdots + \{x_k - y_k\}^2}, \tag{6.1}$$

using Pearson correlation the distance between gene X and Y can be formulated as

$$C_{X,Y} = 1 - P_{X,Y} \tag{6.2}$$

where $P_{X,Y}$ represents the centered Pearson correlation and is defined as

$$P_{X,Y} = \frac{1}{k} \sum_{i=1}^{k} \left(\frac{x_i - \overline{X}}{\sigma_X} \right) \left(\frac{y_i - \overline{Y}}{\sigma_Y} \right) \tag{6.3}$$

where $\overline{X}$ and σ_X are the mean and standard deviation of the gene X, respectively.

The Pearson correlation has value between -1 and 1, with 1 indicating a linear relationship between the two vectors. The Manhattan (or Minkowski) distance [25] between gene X and gene Y is

$$M_{X,Y} = \sum_{i=1}^{k} |x_i - y_i|. \tag{6.4}$$

6.2.3 Gene Clustering and Ordering Using Gene Expression

Hierarchical clustering does not determine unique clusters. Thus the user has to determine which of the subtrees are clusters and which subtrees are only a part of a bigger cluster. So in the framework of hierarchical clustering a gene ordering algorithm helps the user to identify clusters by means of visual display and to interpret the data [5]. For partitive clustering based approaches, clusters are identified by the algorithm automatically and the solutions are robust and not sensible to noise like hierarchical clustering [50]. However, the relationships among the genes in a particular cluster generated by partitive clustering algorithms are generally lost. This relationship (closer or distant) among genes within clusters can be obtained using gene ordering approaches. Ideally, one would like to obtain a linear order of all genes that puts similar genes close to each other; such that for any two consecutive genes the distance between them is small. The gene ordering problem is similar to the Genetic Algorithm based Traveling Salesman Problem solver [34] where, cities are ordered instead of genes [8, 40, 52]. An optimal gene order can be obtained by minimizing the summation of gene expression distances (or maximizing summation of gene expression similarities) between pairs of adjacent genes in a linear ordering $1, 2, \ldots, n$.

A hybrid method (first clustering then ordering) for ordering genes for a hierarchical clustering solution is described in [5]. In [11] the MGO problem is tackled in a memetic algorithm framework where representations and solutions take some ideas from the hierarchical clustering. The impact of different fitness function on the solution is also analyzed in [11]. The ultimate goal is to identify or predict the function of genes from ordering or clustering solutions.

6.2.4 Integrating Other Data Sources with Gene Expression

The challenges of large-scale functional genomics projects are to build a comprehensive map of the cell, protein interactions, gene interactions, and gene functions. The value of combining multi-source information with microarray data, for gene function prediction, is first illustrated in [30]. Consequently, several studies have been reported in recent years which show that combining information from different heterogeneous biological data-sources can enhance the performance over individual data-sources for above mentioned tasks [12, 26, 30, 31, 51]. In [30], proteins are grouped by correlated evolution [37], correlated gene expression [14] and patterns of domain fusion [30] to determine functional relationships among the 6,217 proteins of the yeast Saccharomyces cerevisiae. Many potential protein functions, based on a heuristic combination of these data-sets, are predicted, where, confidence levels for protein-protein links are defined subjectively on a case-by-case basis. Over 93,000 pairwise links between functionally related yeast proteins are discovered and a general function to more than 1000 previously uncharacterized yeast proteins is assigned. Von Mering et al. [31] first developed quantitative methods to measure functional relationship among genes by first benchmarking, and then integrating information from gene fusion [15], chromosomal proximity and phylogenetic profiles [20]. They constructed a global network of functionally interacting proteins, predicted functional modules by using a clustering algorithm, and showed that the predicted modules help in annotating previously uncharacterized proteins and cellular systems. Troyanskaya et al. [51] integrated data sources in Bayesian network approach and predicted functional modules by using a clustering algorithm based on the principle of *KNN* algorithm. The network is constructed on some independence assumptions about different data sources and uses conditional probability tables based on information elicited from yeast experts. Lee et al. [26] derived log likelihood scores from the various data sets, weighted them with a rank-order dependent weighting scheme and added them to find a combined similarity using Bayesian Score. They used a single free parameter for estimating weights and pointed out that a heuristic modification to the strict Bayesian approach performs better for integrating the diverse functional linkage data sets by incorporating the relative weighting of the data. Interacting networks are predicted in [48] which not only identifies highly interacting and functionally connected genes, but also those which are sparsely connected with others. In general, all the methods have a unified scoring scheme, based on gene annotation, for testing the heterogeneous data sets, even when the data sets are accompanied by their own intrinsic scoring method (such as Pearson correlation for gene expression). This re-scoring by a single criterion allows one to directly measure the relative merit of each data set, and then to integrate the data sets.

As mentioned earlier, the present chapter deals with the issues where functions of unclassified genes are predicted by applying clustering and ordering techniques on microarray dataset [39] and by integrating information from different biological data sources [41]. In this regard, first we discuss the characteristics of different data-sources. Two methodologies for predicting gene function, using single and multiple data sources, are described next along with experimental results.

6.3 Data Sources

The availability of high-throughput biological data has opened a new direction in genomic analysis and function prediction of unclassified genes in recent years. The challenge is to find a efficient way to integrate these increasing rich trove of evidence to infer biological function. Many of these data, such as protein localization, structure, function and expression, post-translational modifications, molecular and genetic interactions, phenotypic profiles [9], gene expression microarrays [14], protein sequences [30], Kyoto Encyclopedia of Genes and Genomes (KEGG) pathway information [23], protein-protein interaction data [42, 43], protein phytogenetic profiles [37] and Rosetta Stone sequence [29] information assess functional relationships between genes on a large scale and can be the key to assign functional annotation to a significant number of unclassified genes [51]. Some of this broad set of functional data has been already assembled for function prediction of unclassified genes. Now we will describe the details of different data sources and their respective pair-wise gene similarity extraction techniques.

Phenotypic Profile

Phenotypic profile deals with the quantitative sensitivity profiles of large number of strains with individual homozygous deletion of all nonessential genes [9]. Each gene is replaced by a cassette containing a 20-mer molecular 'barcode' unique for each deletion mutant. Brown et al. [9] first used phenotypic profiles for global analysis of the function of genes in budding yeast by hierarchical clustering of the quantitative sensitivity profiles of the 4756 strains. The clustering solutions are then used for identifying function of genes involved in various DNA repair, damage checkpoint pathways, and other interrogated functions. Analysis of the phenotypic profiles places a total of 860 genes of unknown function in clusters with genes of known function. This complete phenotypic profile data is used here as one of the data sources. Pearson correlation is used as a similarity extraction technique for phenotypic profile [9].

Let $X = x_1, x_2, \ldots, x_k$ and $Y = y_1, y_2, \ldots, y_k$ be the phenotypic profiles of two genes obtained over a series of k different treatments. Using centered Pearson correlation, the similarity between genes X and Y is defined as

$$Pc_{X,Y} = \frac{1}{k} \sum_{i=1}^{k} \left(\frac{x_i - \overline{X}}{\sigma_X} \right) \left(\frac{y_i - \overline{Y}}{\sigma_Y} \right) \tag{6.5}$$

where $\overline{X}$ and σ_X are the mean and standard deviation of the gene X, respectively. σ_X is defined as

$$\sigma_X = \sqrt{\frac{1}{k} \sum_{i=1}^{k} (x_i - \overline{X})^2}. \tag{6.6}$$

The Pearson correlation has value between −1 and 1, where 1 indicates a linear relationship between the two vectors.

When the phenotypic profile data set is downloaded from the website it is found that out of 51 treatments some treatments are missing for some genes (strains). For the subsequent analysis to be as informative as possible, it is essential that the missing values have to be estimated in order to analyze the available data and the coefficients for the missing values are as accurate as possible. Currently, there is no state-of-the-art missing value estimation method for phenotypic profiles. Alternatively, missing values in phenotypic profiles can be estimated using the methods that are used for microarray gene expression. In this phenotypic profile data set, all the genes with more than 50% missing values are first eliminated from the dataset. Thereafter for the remaining genes missing values are estimated using LSimpute [18] software, a state-of-the-art statistical java based package to estimate missing values in gene expression data set. In LSimpute software, two basic methods based on least squares principle, one utilizing correlations between genes (LSimpute_gene) and the other utilizing correlations between arrays (LSimpute_array), are used to estimate missing values. A robust method (LSimpute_adaptive) using weighted averages of the estimates from LSimpute_gene and LSimpute_array for adaptive estimates is also available in LSimpute.

Phenotypic profile data is susceptible to biases created during the PCR amplification reaction. The detailed procedure of generating and normalizing the data is available in Brown et al. [9]. In brief, each gene (strain) in phenotypic profile data is associated with four hybridization signals on the high-density oligonucleotide array generated in two separate PCR labeling reactions [9]. The data is then normalized by Brown et al. in the experimental array to that of the control array in order to eliminate any bias created during the PCR amplification reaction. The normalized data is downloaded from the supplementary material of [9].

Gene Expression

To identify relationship among genes, involved in multiple biological functions or processes, many microarray experiments [14, 47] with different biological origins are conducted world wide. While, array technologies have made it easy to monitor simultaneously the expression patterns of thousands of genes during cellular differentiation and response, the challenge still remains in meaningful analysis of such massive data sets. For simple experiments with just two samples, it is enough to rank the genes by their relative induction. However, complex experimental designs, like microarray, could involve thousand of genes and hundreds of samples for example, complete developmental time courses in many cell lines. No two genes are likely to exhibit precisely the same response, and many distinct types of behavior may be present. Moreover, microarray experiments with multiple slides are also sources of non-biological variation between slides such as dye biases, sample preparation or hybridization differences, scanner calibrations, slide printing variations, volume of initial RNA, etc. Some of these variabilities can be corrected by "data normalization" before analysis of the data. Normalization can be performed by removing saturated signals, background correction, low expression genes correction, etc. and many different methods are developed in order to compensate for dye-effects and other non-biological variations between arrays [57].

A key goal of microarray experiments is to extract the fundamental patterns of gene expression inherent in the data. Here, the All Yeast data [14, 54], having 6221 genes and 80 time points, is used for microarray gene expression analysis. The All Yeast dataset is downloaded from Stanford Microarray Database [45] with default parameters. The default parameters include all types of normalization applicable to that data and suggested by the experts. A number of measures can be used in finding the microarray gene expression similarity for gene annotation and grouping. The most popular and probably most simple measures for finding global similarity between genes are the Pearson correlation [51], a statistical measure of (linear) dependence between random variables, and the Euclidean distance [30]. Centered Pearson correlation is used for extracting gene expression similarity.

Microarray experiments often produce multiple missing expression values, normally due to various experimental problems. As gene expression analysis generally requires a complete data matrix as input, the missing values have to be estimated in order to analyze the available data. Alternatively, genes with missing expression values can be removed until no missing values remain. However, for arrays with only a small number of missing values, it is desirable to estimate those values [18]. For the All yeast data, we estimated the missing values using LSimpute_adaptive [18].

KEGG Pathway

The Kyoto Encyclopedia of Genes and Genomes (KEGG) database [23] provides pathway information for genes involved in 221 pathways related to metabolism, genetic information processing, environmental information processing and cellular processes. This information can be used as a reference for functional reconstruction. For each of these 221 pathways, all the protein sequences except yeast proteins, are downloaded from PIR [6]. Profile vector for each protein in yeast is computed by comparing its sequence across 221 pathway databases, using BLAST [1]. The method is similar to phylogenetic profile [37] construction, where, each pathway database is replaced by all proteins within a species. The pathway profiles of genes, computed using KEGG pathway databases, are denoted as KEGG profiles.

To find the similarity between two genes using KEGG profiles, the ratio of dot product value and OR value between two profiles is used. The similarity matrix has a highest similarity value of 1. Hence, the similarity values, obtained by all pair-wise comparison, have a dynamic range from 0 to 1 and its normalization is unnecessary.

Protein Sequence

The historical method of finding the function of a protein involves extensive genetic and biochemical analyses, unless the amino-acid sequence of the protein resembles another whose function is known. Comparative analysis of protein sequences is a prominent and crucial approach for analysis, annotation and deciphering functions of genes and proteins. But, protein similarity information contains mostly known and validated protein (or gene) relations. Intuitively one can assume that all the protein relations arising from direct protein similarity search is available in the literature and will not help in predicting functions for unclassified genes in a widely studied organism like yeast. As compared to direct protein similarity search, the field of searching gene/protein similarity through phylogenetic profiles (PP) [37], Rosetta Stone sequence (RS) [29], and transitive homology [35] are relatively new methods and with increasing number of fully sequenced genomes the search space of these methods are increasing rapidly.

The accuracy of transitive homologues for extracting protein similarity is higher than related methods as reported in [28, 56]. The method has been shown to be better than a direct pairwise homology search [35]. Transitive homology detection method [28, 35, 56] works by searching the query sequence against the database with a conservative threshold to find the closely homologous sequences and using these homologous sequences as seeds to search the database to find remotely homologous sequences with a less conservative threshold. For example, consider the transitive homology between sequence a and sequence b through the third sequence c. The E-values between sequence a and sequence c, sequence c and sequence b, as well as sequence a and sequence b are 0.01, 0.005, and 20 respectively. The protein similarities $B_{a,c}$, $B_{c,b}$, and $B_{a,b}$ are 0.8, 0.9, and 0.2 respectively. The homology between sequence a and sequence b cannot be detected with their direct E-value. However, the value of $B_{a,b}$ is assigned to $0.8 \times 0.9 = 0.72$ because of the transitive sequence homology. This transformation takes advantage of the transitive homology of sequences A and B through the intermediate sequence C, assuming that sequences A and C and sequences B and C are independently homologous [28]. To find the transitive homologues, homology comparisons are performed among target proteins and 37,66,477 proteins downloaded from UniProt [4], by using BLASTP in BLAST [1]. Instead of storing raw BLAST score as the similarity between two protein sequences, the metric of ProClust [38] is used where the metric value scales from 0 to 1. It is the ratio of the raw BLAST score of the sequence alignments to the raw BLAST score of one of those two sequences aligned to itself.

The transitive homology based method may provide false positives due to domain shuffling events and can consider homology where, only small regions of the proteins (like domains) are conserved. In this regard, a threshold of 200 residues is considered on total length of the protein alignment as the size of majority (90%) of domains is less than 200 residues.

Protein-Protein Interaction

Protein interactions assemble the molecular machines of the cell and represent the dynamics of virtually all cellular responses [36]. While, genetic interactions reveal functional relationships between and within regulatory modules, protein-protein maps reveal many aspects of the complex regulatory network underlying cellular function [17, 31]. Such interactions along with protein-protein maps, defines the global regulatory network of the cell. In this regard, biological general repository for interaction datasets have been developed by various groups to store and distribute comprehensive collections of physical and genetic interactions. In this study, manually curated catalogues of known protein-protein interactions are downloaded from BioGRID [49] on September 2010 and binary interactions are used as the common unit of analysis. For a given pair of genes/proteins the similarity value is 1 or 0, indicating a interaction present or absent, respectively. Since the similarity value scales from 0 to 1, its normalization is unnecessary. The BioGRID database/catalogue includes 2,38,313 interactions for yeast genes by combining results obtained from synthetic lethality, affinity capture, two-hybrid, epistatic miniarray profile, reconstituted complex, co-crystal structure, co-purification, dosage rescue, phenotypic enhancement, phenotypic suppression, synthetic growth defect, co-fractionation, biochemical activity, synthetic rescue, and protein-peptide based experiments. The related references are also available in BioGRID.

6.3.1 Evaluation for Dependence Among Data Sources

A multi-source integration framework is meaningful, when these sources are independent of each other. In this regard, the statistical dependence of every data source with respect to other data sources, using distance correlation, is checked in [41]. Distance correlation is a measure of statistical dependence between two random variables or two random vectors of arbitrary, not necessarily equal dimension. Its important property is that this measure of dependence is zero if and only if the random variables are statistically independent. Its maximum value is 1, indicating an absolute dependence. The distance correlation between two variables, X and Y, is defined as

$$dCor(X, Y) = \frac{dCov(X, Y)}{\sqrt{dVar(X)dVar(Y)}} \qquad (6.7)$$

where, $dCov(X,Y)$ is the distance covariance between X and Y and $dVar(X)$ and $dVar(Y)$ are the distance variances of X and Y, respectively.

Table 6.1 summarizes the results obtained by calculating the distance correlations among different data sources. Here, PhP, GE, KP, TH, and PPI indicates phenotypic profile, gene expression, KEGG pathway, transitive homology, and protein-protein interaction, respectively. From the results it is observed that the off diagonal elements

Table 6.1 Distance correlations among data sources

Source	PhP	GE	KP	TH	PPI
PhP	1.00	0.09	0.11	0.10	0.00
GE	0.09	1.00	0.03	0.07	0.00
KP	0.11	0.03	1.00	0.19	0.00
TH	0.10	0.07	0.19	1.00	0.00
PPI	0.00	0.00	0.00	0.00	1.00

are close to 0.1 or 0.0, and indicate that the dependence is negligible among data sources.

6.3.2   Relevance of Data Sources

Gene annotation [3, 13] information is one of the most important biological knowledge that is currently used for validation of clustering and classification results recovered by data analysis and also for evaluating the relevance of data sources. Most of the gene annotation information is in the form of genes grouped by the biological function, process, component or sub-cellular location [13]. Result interpretation steps can also be enhanced by a broad variety of annotation information about genes under study, including pathways, transcription factors, chromosomal location, etc.

Five data sets are used to integrate information and to predict gene function. This data sources are accompanied by their own way of extracting gene similarities (such as Pearson Correlation for phenotypic profile) and they have different observational scales. One can evaluate the relevance of each data source if a unified observational scale can be identified. In this regard, re-scoring the similarities in the common scale of positive predictive value (*PPV*) using yeast GO-Slim process annotations of genes in the SGD database [13] can provide a unified framework. Note that, according to yeast GO-Slim process and MIPS, there are 6069 and 6130 annotated genes (ORFs) for yeast of which 4387 and 4737 genes, respectively, are classified to some biological or functional process and the remaining genes are unclassified. The *PPV* at a particular similarity value (computed from a data source), using gene annotations, is defined as [51]

$$PPV = \frac{\textit{no. of gene pairs sharing common annotations}}{\textit{total no. of gene pairs}}. \tag{6.8}$$

where, gene pairs sharing common annotations are pairs of genes having an overlapping yeast GO-Slim process annotation and the total no. of gene pairs is the available gene pairs at a particular similarity value for a particular data source. To obtain the relevance of data sources, Ray et al. [41] computed the *PPV*s at various similarity values for each of them and plotted them in the same figure. Such a relevance

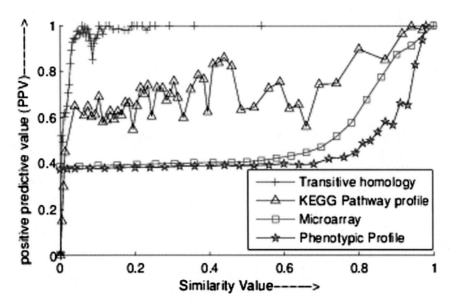

Fig. 6.1 Comparing the re-scored similarity values for different types of data sources to obtain equivalency in the common framework of Yeast GO-Slim process annotations. The positive predictive values (*PPV*) versus the similarity values are plotted for each data source

comparison is available at Fig. 6.1. The protein-protein interactions (not shown in Fig. 6.1) are binary relations and have a constant *PPV* of 0.69 at a similarity value of 1. From the figure it can be observed that protein similarity through transitive homology is the most relevant one in terms of *PPV* and it lies at the top of all other curves. The relevance of phenotypic profile is the lowest but still enough to include it as a datasource as many points have *PPV* above 0.50.

6.4 Gene Function Prediction

We mainly focussed on yeast (Saccharomyces Cerevisiae), a model eucaryotic organism, for gene function prediction. According to SGD and MIPS, there are more than 6000 annotated genes (ORFs) for yeast of which 4387 and 4737 genes, respectively, are classified to some biological or functional process and the remaining genes are unclassified. As mentioned in before, the availability of high-throughput biological data like microarray has opened a new direction in function prediction of unclassified genes in recent years. Further, this type of data can help in establishing the functional relationships between genes on a large scale and can be the key to assign functional annotation to a significant number of unclassified genes. However, the sheer volume of information of these data makes manual integration, evaluation and function prediction impracticable. Therefore, automated algorithms have been developed to accomplish the above mentioned tasks.

6.4.1 Prediction Using Single Data Source

As mentioned earlier, relationship among the various data points (here genes) is generally lost in individual clusters obtained using any partitive clustering method. However, it can be constructed by ordering the data points in each cluster. Here we describe the hybrid method developed in [39] for ordering genes in each of the clusters obtained from partitive clustering solution, using microarray gene expressions. In this hybrid approach an algorithm for optimally ordering cities in traveling salesman problem (TSP), called FRAG_GALK [40], is hybridized separately with CLICK [44], k-means [19] and self organizing map (SOM) [50] to show the importance of gene ordering in partitive clustering framework for analyzing gene expressions. First, the solutions are obtained using clustering algorithm and then the ordering technique of FRAG_GALK is used on each cluster. Here we briefly discuss the various steps used in FRAG_GALK, which are also available in [40]. The steps are:

Step 1: Create the string representation (chromosome of GA) for a gene order (an array of n integers), which is a permutation of 1, 2, ..., n with nearest-neighbor (NN) heuristic. Repeat this step to form the initial population of GA.

Step 2: Apply Nearest Fragment (NF) heuristic on each chromosome probabilistically.

Step 3: Each chromosome is upgraded to local optimal solution using chained LK heuristic [2] probabilistically.

Step 4: Fitness of the entire population is evaluated and elitism is used, so that the fittest string among the child population and the parent population is passed into the child population.

Step 5: Using the evaluated fitness of entire population, linear normalized selection procedure is used.

Step 6: Chromosomes are now distributed randomly and modified order crossover operator is applied between two consecutive chromosomes probabilistically.

Step 7: Simple inversion mutation (SIM) is performed on each string probabilistically.

Step 8: Generation count of GA is incremented and if it is less than the maximum number of generations (predefined) then from step 2 to step 6 are repeated.

The final solution obtained by the hybrid technique is evaluated in terms of gene expression distances as [39]

$$F_1(n) = \sum_{j=1}^{k} \sum_{i=1}^{N_j-1} C_{i,i+1}^{j}, \qquad (6.9)$$

where k is the total number of clusters, N_j is the number of genes in cluster j, and $C_{i,i+1}^j$ is the distance/similarity between two adjacent genes i and $i + 1$ in cluster j.

A score, that is different from the similarity/distance measure and defined in Eq. 6.10, is used to evaluate the final gene ordering. Each gene that has undergone MIPS categorization can belong to one or more categories, while there are many unclassified genes also (no category). A vector $V(g) = (v_1, v_2, \ldots, v_j)$ is used to represent the category status of each gene g, where j is the number of categories. The value of v_j is 1 if gene g is in the jth category; otherwise is zero. Based on the information about categorization, the score of a gene order for multiple class genes is defined as [52]

$$S = \sum_{i=1}^{N-1} G(g_i, g_{i+1}), \tag{6.10}$$

where N is the number of genes, g_i and g_{i+1} are the adjacent genes and $G(g_i, g_{i+1})$ is defined as

$$G(g_i, g_{i+1}) = \sum_{k=1}^{j} V(g_i)_k V(g_{i+1})_k, \tag{6.11}$$

where $V(g_i)_k$ represents the kth entry of vector $V(g_i)$. For example consider the gene order

g_1, g_2, g_3

with 15 categories represented by vectors

$$V(g_1) = (1, 0, 1, 1, 0, 0, 0, 0, 0, 0, 0, 0, 0, 0, 0) \tag{6.12}$$

$$V(g_2) = (1, 1, 1, 1, 0, 0, 0, 0, 0, 0, 0, 0, 0, 0, 0) \; and \tag{6.13}$$

$$V(g_3) = (0, 0, 1, 0, 0, 0, 0, 0, 1, 0, 0, 0, 0, 0, 0) \tag{6.14}$$

Then $G(g_1, g_2) = 3$ and $G(g_2, g_3) = 1$, and

$S(n) = G(g_1, g_2) + G(g_2, g_3) = 3 + 1 = 4$.

Note that, S can be used as scoring function for single class genes also. Using scoring function S, a gene ordering would have a higher score when more genes within the same group are aligned next to each other. So higher values of S are better and can be used to evaluate the goodness of a particular gene order.

6.4.2 Results and Biological Interpretation

Performance of the FRAG_GALK for gene ordering is compared mainly with Concorde_LP [2] and Bar-Joseph et al.'s [5] method. Experiments of gene ordering are conducted in Matlab 7 on Sun Fire V 890 (1.2 GHz and 8 GB RAM). The codes for B-Joseph et al.'s [5] leaf ordering in hierarchical clustering solution are downloaded from [53]. Finally, FRAG_GALK and Concorde_LP are applied separately on the gene clusters obtained by SOM, and B-Joseph et al.'s method is applied on the average linkage based hierarchical clustering solution for each dataset. While FRAG_GALK is a genetic algorithm (GA) based TSP solver, Concorde is a linear programming based TSP solver and much slower than FRAG_GALK. Note that, SOM and K-means are available in Expander [44] and used with 16, 16, and 18 clusters for clustering Cell Cycle, Yeast Complex, and All Yeast data sets respectively as genes in these datasets are classified according to MIPS [32] categorization into 16, 16, and 18 groups. The MIPS (Munich Information Center for Protein Sequences) [32, 33] categorization is available for all the three Yeast data sets that allows a gene to belong to more than one category.

The ultimate goal of an ordering algorithm is to order the genes in a way that is biologically meaningful. In this regard, Table 6.2 compares the performance of the hybrid approach, using FRAG_GALK, with Concorde and B-Joseph's [5] leaf ordering in hierarchical clustering solution in terms of the $F_1(n)$ value (Eq. 6.9), S value (Eq. 6.10), and computation time. From the biological scores (Table 6.2), it is evident that FRAG_GALK provides biologically comparable gene order with respect to Concorde and sometimes superior gene order than 'leaf ordering in hierarchical clustering solution' by [5], for all datasets in least computational time. For example, FRAG_GALK took 125 s to order All Yeast data (6221 genes) as compared to Concorde and B-Joseph et al.'s method which took 2272 and 1989 s respectively.

Table 6.2 Summation of gene expression distances ($F_1(n)$), biological Score (S), and computation time of ordering in seconds (within parenthesis) for different algorithms

Algorithm	Data sets		
	Cell Cycle	Yeast Complex	All Yeast
SOM	442.94	547.16	3446.60
	354	792	1730
SOM+FRAG_GALK	301.72	330.54	1919.15
	386	1011	2356
	(0.7)	(1.13)	(125)
SOM+Concorde	301.72	330.54	1919.15
	386	1011	2356
	(3.41)	(15.26)	(2272)
B-Joseph	300.51	330.17	1920.82
	381	1024	2350
	(1.8)	(3.34)	(1989)

To show the utility of the hybrid method in identifying different subclusters within big clusters and grouping the functionally correlated genes within clusters for Yeast Complex data, here for illustration, gene subclusters found by SOM+FRAG_GALK and their functional category indexes in first 6 clusters are provided in Table 6.3. The functional categories corresponding to different indexes are shown in Table 6.4. Yeast Complex data is first clustered in 16 groups using SOM, but the subclusters of highly coregulated genes, as shown in Table 6.3, could not be identified if SOM is used alone. For example, all the 9 genes in the 3rd subcluster of cluster 4 (YBR010W, YNL031C, YBL003C, YDR225W, YDR224C, YNL030W, YBR009C, YBL002W and YPL256C) are involved in Cell Cycle and DNA processing, Transcription, and Protein with Binding Function or Cofactor Requirement. While using SOM these 9 genes are distributed in the cluster 4, using CLICK these 9 genes are not assigned to any cluster, and are left as singletons along with 263 other singleton genes, which is undesirable as they belong to the same biological categories. After ordering genes using FRAG_GALK in cluster 4 of SOM and singleton genes of CLICK they (the 9 genes) are tightly grouped and identified easily.

Table 6.3 Gene subclusters found by SOM+FRAG_GALK and their functional category indexes in first 6 clusters identified using SOM for Yeast Complex data

Cluster	Subcluster	Genes	Functional index
1	Nil	100 genes	5
2	1	YGR162W, YBR079C, YMR309C, YOR361C, YNL062C, YHR021C, YBL092W, YBR143C, YDL136W, YDL191W, YOR167C, YJL191W	
	2	YOR224C, YJR063W, YNL113W, YBR154C, YGL120C, YNL248C, YJL148W, YOR340C, YPR110C	4 and 7
	3	YMR146C, YOL139C, YOR276W, YKL156W, YOR182C, YDR447C, YKR094C, YIL148W, YDR500C, YMR121C, YLR264W, YJL138C, YKR059W, YKL081W, YLR249W	5
	4	YGR159C, YOR207C, YMR239C, YHR089C, YOR310C, YLR197W, YLL008W	4
3	1	YMR260C, YDR429C, YPL237W, YLR406C, YJR007W, YER025W, YPR041W, YDR172W, YDR211W	5
	2	YDR212W, YIL142W, YPL210C, YKL057C, YPL243W	6 and 7
	3	YLR060W, YOR260W, YDL040C, YKR026C, YLR291C, YBR142W, YBL087C, YHL001W, YDR450W, YHL033C, YBR191W, YBR189W, YBR048W, YBR118W	5
	4	YBR142W, YHR062C, YHR065C, YNR003C, YMR043W, YIL021W, YOR210W, YDR194C, YHR069C	4

(continued)

Table 6.3 (continued)

Cluster	Subcluster	Genes	Functional index
4	1	YLR093C, YNL121C, YLR170C, YML112W, YBR160W, YBR171W, YLR378C, YML019W, YPL234C, YOR039W	6
	2	YKR068C, YLL050C, YGL200C, YML012W, YPL218W, YKL080W, YDR086C, YNL153C, YKL122C, YLR292C, YGL112C, YLR268W, YLR447C	6 and 9
	3	YBR010W, YNL031C, YBL003C, YDR225W, YDR224C, YNL030W, YBR009C, YBL002W, YPL256C,	3, 4, and 7
	4	YJL025W, YPR101W, YMR061W, YGR195W, YOR244W, YLR105C, YDL043C, YPR056W, YPR057W	4
	5	YGL100W, YNL261W, YKL144C, YNL151C, YJL008C, YER148W	7
5	1	YML051W, YNL103W, YCR035C, YJL006C, YOL135C, YDL165W, YOL102C, YMR270C, YOR085W, YJL002C, YKL211C	1 or 4 or (1 and 4)
	2	YBR087W, YJR043C, YJL173C, YJR076C, YMR080C, YPR029C, YKL045W, YLR103C, YNL262W, YKL113C, YLR212C, YER070W, YIL066C, YDR097C, YOL090W, YDL102W, YOR250C	3 or (3 and 7) or (1, 3 and 7)
6	Nil	49 genes	3

Table 6.4 Indexes and corresponding functional category

Functional index	Functional category
1	Metabolism
2	Energy
3	Cell cycle and DNA processing
4	Transcription
5	Protein synthesis
6	Protein fate (folding, modification, destination)
7	Protein with binding function or cofactor requirement
8	Protein activity regulation
9	Cellular transport, transport facilitation and transport routes
10	Cellular communication/signal transduction mechanism
11	Cell rescue, defense and virulence
12	Interaction with cellular environment
13	Cell fate
14	Development (systemic)
15	Biogenesis of cellular component
16	Cell type differentiation

Let us consider another typical example where Herpes data is clustered using 5 groups as there are 5 categories [21], namely, Homologs of cellular regulatory or signal transduction genes, Virus gene regulation, DNA replication, DNA repair and nucleotide metabolism, and Virion formation and structure [21]. In this classification one gene can belong to only one category. The CLICK algorithm clustered Herpes data in one group (Fig. 6.2a) of 101 genes and 5 singleton genes and no subcluster can be identified within it. When the genes are ordered in the cluster using FRAG_GALK, some subclusters are identified using visual display (Fig. 6.2b), but the display is not conclusive in identifying exact number of sub-clusters. Using K-means, Herpes data is clustered with k = 5 (5 clusters). Visual display of the first four clusters are shown in Fig. 6.2c. After ordering the genes in each cluster using FRAG_GALK, 3 and 3 distinct patterns are observed in cluster 1 and 2 respectively (Fig. 6.2d). From Fig. 6.2e it is found that B-Joseph's method is not very efficient (genes with all red expressions are scattered through out the visual display) in identifying clusters as partitive clustering algorithms. On the other hand, hybridizing FRAG_GALK with CLICK (Fig. 6.2b) or K-means (Fig. 6.2d) provided a robust clustering method as well as relations between genes within clusters. With all these ordered and clustered genes one can easily zoom in a useful small subset of genes in a cluster which cannot be done alone with partitive clustering methods. In a similar way, subclusters within big clusters are identified by Concorde_LP for all the data sets.

6.4.3 Prediction Using Multiple Data Sources

Even though predictions of genetic interactions within the cell are still far from being correct, computational methods has begun to forge connections between cellular events and to foster numerous hypotheses on gene functions. While gene expressions and phenotypic profiles are excellent tools for hypothesis generation, they alone often lack the degree of specificity needed for accurate gene function prediction. This improvement in specificity can be achieved through the incorporation of heterogeneous functional data in an integrated analysis [51]. In this regard, the scoring of different data sources, based on the unified observational scale of *PPV*'s, allows us to directly compare and integrate the different types of datasets. The *PPV* reflects the accuracy of data sets, but do not provide any information about relative weight estimate of one data source in presence of the other data sources. Here we describe the method in [41] which is focussed on integrating information from data sources in a weighted arithmetic mean framework. Functional categories of 12 unclassified yeast genes are also predicted.

A 'Biological Score' score, called *BS* is developed, where, *PPV* computed from phenotypic similarity (*Pp*), gene expression similarity (*Pm*), KEGG pathway profile similarity (*Kp*), protein similarity through transitive homologue (*B*), and protein-protein interaction information (*I*) between two genes *X* and *Y* are integrated through weights *a*, *b*, *c*, *d*, and *e* in a linear combination style. This score is referred to as *Biological Score* (*BS*) and is defined as

(a)   (b)   (c)   (d)   (e)

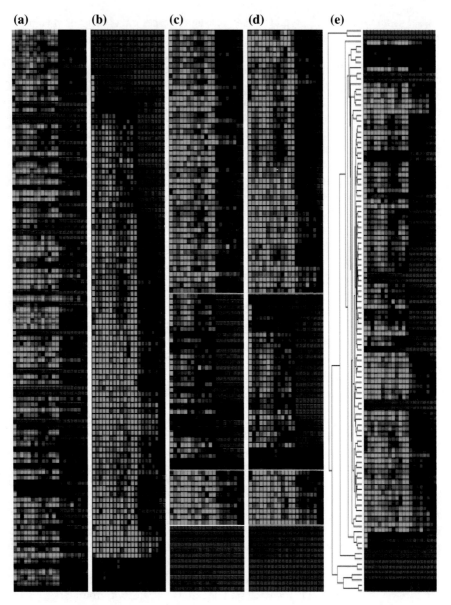

Fig. 6.2 Comparing **a** CLICK, **b** 'CLICK + FRAG_GALK', **c** *K*-means, **d** '*K*-means+ FRAG_GALK', and **e** B-Joseph's method for Herpes data

$$BS_{X,Y} = \frac{a \times Pp_{X,Y} + b \times Pm_{X,Y} + c \times Kp_{X,Y} + d \times B_{X,Y} + e \times I_{X,Y}}{a + b + c + d + e} \quad (6.15)$$

where a, b, c, d, and e are varied within range 0 to 100 in steps of 0.1 to find a combination that maximizes the *PPV* for a user defined number of top gene pairs. Note that, the weights a, b, c, d, and e are assigned to the complete *PPV* matrices calculated from individual data sources. The following can be stated about the score:

1. $0 \leq BS_{X,Y} \leq 1$
2. $BS_{X,Y} = BS_{Y,X}$ (symmetric).

The scoring framework for data source integration, in Eq. 6.15, is based on data source weighting where the re-scored similarity spaces, available from different data sources, are adaptively transformed using a set of weighting coefficients. Intuitively, more important similarity spaces should be assigned larger weights than less important ones, while irrelevant ones should be assigned zero weight. The following steps are used to estimate the weight factors a, b, c, d, and e in the *Biological Score*:

Step (1) All the factors are assigned an initial value of 1.
Step (2) *BS* values are calculated for all the gene pairs and sorted in descending order to identify the gold standard cut-off value above which the top 26,432 gene pairs are available. The gold standard cut-off value and gold standard of top gene pairs are determined from KEGG pathway profiles, which provides 26,432 gene pairs with similarity value 1 and constant *PPV* of .81. These gene pairs are the most predictive of all, whereas the *PPV* of other data sources, as well as gene pairs below top 26,432 for KEGG pathway profiles, vary considerably.
Step (3) *PPV* is calculated for the top 26,432 gene pairs.
Step (4) The weight factors are now varied in steps of 0.1 and the steps from 2 to 3 are repeated to find a combination of weights for which the *PPV* is maximized.

Some typical instances in the exhaustive search process, showing the variation of *PPV* of *BS* for different sets of weight coefficients (assigned to *PPV*s of different data sources), ranging from 0 to 100, in steps of 1, are shown in Fig. 6.3. Yeast GO-Slim process annotations are used in determining the *PPV*s.

Results and Biological Interpretation

The function for each gene is predicted from the top K neighbors and selecting a gold standard *BS* cut-off value obtained from KEGG pathway profiles using MIPS classification. The gene clustering method is denoted as *K-BS*, where each gene is considered once for its function prediction and allows its neighbor genes to be a member of multiple gene clusters. This clustering method, based on K nearest neighbors of each gene with K = 10, is already used in previous related investigations of Marcotte et al. [30] and Troyanskaya et al. [51]. As Yeast GO-Slim process

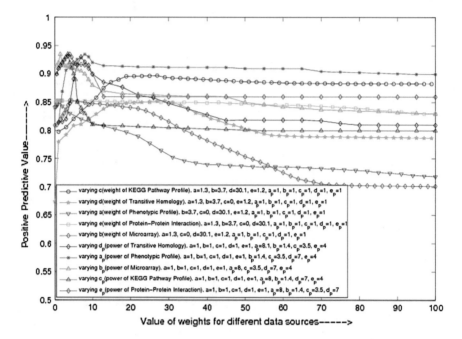

Fig. 6.3 Comparing the values of *PPV* using *BS*, by varying weight coefficients of *PPV* of different data sources for top 24632 gene pairs. When a particular coefficient is varied the others are kept constant at the values shown in the figure

annotations was used for determining the weights of the data sources, 510 different MIPS functional categories are used to evaluate the biological significance of the clusters generated by *K-BS*. One or several predominant functions are then assigned to each cluster and the target gene (the gene whose K nearest neighbors, using *BS* as a similarity value, are considered to form the cluster) by calculating the P-values for different functional categories. Several clusters are observed to be significantly enriched with genes having similar function. The top predictions consist the function of 12 unclassified (MIPS 2007) and 417 classified genes at *BS* cut-off value 0.77, and *P*-value cut-off 1×10^{-13}. At these cut-off values, the functions of the classified genes are predicted with 98.20 *PPV*. Table 6.5 summarizes the top 12 predicted functions for 12 unclassified genes. For each of the predicted functions, the related p-values, no. of related genes in the cluster and the genome, are also shown in the table.

Since four out of the 12 clusters show functional enrichment in a single category of 'DNA topology', these clusters are manually analyzed. It is observe that 15 classified (YBL113c, YDR545w, YEL077c, YER190w, YGR296w, YHL050c, YIL177c, YJL225c, YLL066c, YLL067c, YLR466w, YLR467w, YNL339c, YPL283c, and YPR204w), 4 unclassified (YHR218W, YHR219W, YBL112C, and YLR464W) and 2 recently deleted (YEL076C and YPR203W) genes are distributed in these clusters with 80% genes in common. Further, clustering is also performed with *K-BS* by

Table 6.5 Top 12 function predictions of unclassified gene at *BS* cut-off value of 0.77

Unclassified gene	Functional category	*P*-value	Genes within cluster	Genes within category
YIL080W	ABC transporters	2.2204e-16	8	28
YLR057W	Modification with sugar residues	2.2871e-14	8	67
YHR218W	DNA topology	0	9	52
YHR219W	DNA topology	0	10	52
YIL170W	C-compound and carbohydrate transport	1.3656e-14	8	63
YDR441C	Purin nucleotide/ nucleoside/ nucleobase metabolism	6.7724e-15	8	58
YCL074W	Transposable elements, viral and plasmid proteins	3.3307e-16	8	34
YBL112C	DNA topology	0	10	52
YLR464W	DNA topology	2.6645e-15	8	52
YMR010W	Modification with sugar residues (e.g. glycosylation, deglycosylation)	0	9	67
YIL067C	Vesicle fusion	2.2204e-16	9	32
YHR049W	Metabolism of secondary products derived from glycine, L-serine and L-alanine	3.3307e-16	7	19

selecting $K = 20$ to find if these four clusters merge to form a single cluster. After clustering, all the 21 genes are found in the same cluster, which shows functional enrichment in categories 'CELL CYCLE AND DNA PROCESSING' (15 out of 19, *P*-value 1.53×10^{-10}), 'DNA processing' (15 out of 19, *P*-value 7.02×10^{-15}) and 'DNA topology' (15 out of 19, *P*-value 2.38×10^{-30}). Our analysis predicts that the four unclassified genes are very likely to be involved in the above mentioned processes. On examination of the literature for 4 unclassified genes, we it is found that their involvement in DNA processing and DNA topology is likely due to their relation to helicase-proteins [13]. These proteins play important roles in various cellular processes including DNA replication, DNA repair, RNA processing, chromosomal segregation, and maintenance of chromosome stability.

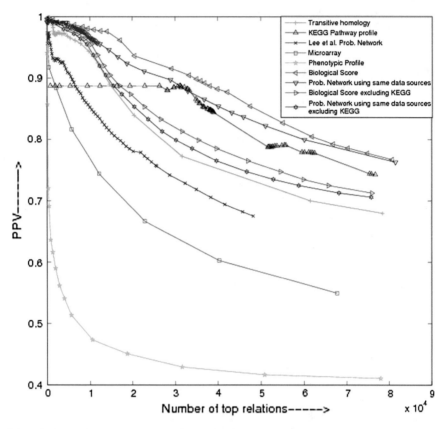

Fig. 6.4 Comparison between the *Biological Score* (*BS*), Lee et al.'s Probabilistic Network, and individual data source in terms of *PPV* versus the number of top gene pairs. While, the available annotations (using vector $V(g)$ as in Eq. 6.12) from Yeast GO-Slim process is used to train the weighting factors in *BS* and 'Probabilistic Network using same data sources', the available annotations from MIPS are used to evaluate (using *PPV*) the gene pairs of all the methods and data sources

In order to demonstrate the power of data source integration, the *PPV* of gene pairs identified by *BS* is compared with those identified by the individual data sources. Since *BS* uses GO annotations for adapting its weights, it is not used for performing the comparisons. Rather, the MIPS annotation of classified genes is used (Fig. 6.4). The similarity values computed from *BS*, phenotypic profiles, gene expression, KEGG profiles, and protein similarity from transitive homology are sorted in descending order, and a curve is drawn for top gene pairs verses *PPV* from the sorted data for each form of data source. In contrast, *PPV* for protein-protein interactions has a constant value of 0.69 and not shown in Fig. 6.4. It is observed that the curve of *BS* is above the other curves. Moreover, the top 26,432 gene pairs has an *PPV* greater than the gold standard KEGG pathway profiles. The gene pairs are also reasonably distinct from gene pairs of KEGG pathway profiles. It demonstrates that *BS* achieved higher *PPV* by combining similarities from multiple sources. Figure 6.4 also compares the performance of *BS* and 'final log likelihood scores' of Lee et al.'s probabilistic network (downloaded from the website mentioned in [27]) in terms of *PPV* with MIPS annotation. The top 1,00,000 gene pairs predicted by *K-BS* method with *PPV* above 0.755 (not shown in the data) are available in http://www.isical. ac.in/~scc/Bioinformatics/AdS/toprelation.txt in tabular (tab delimited) form. The *PPV* computed from individual data source are also shown in the file.

6.5 Relevance of Soft Computing and Granular Networks

The hybrid gene ordering approach using a partitive clustering method and FRAG_ GALK can provide good solutions in less computation time than other methods as genetic algorithm (GA), one of the soft computing tools, is used. GA can augment the search space quickly and helps in obtaining the global optimal solutions. Further, the nearest fragment heuristic in FRAG_GALK reduces the limitation of Nearest Neighbor (NN) heuristic in path construction by determining the optimum number of fragments in terms of the number of genes and then greedily reconnecting them. Although the fragment reconnection is performed on local decisions, the random slicing of chromosomes (in GA) into optimum number of fragments helps FRAG_GALK in obtaining the global optimal solutions to some of the gene ordering problems. It will be worthwhile to test the performance of this hybrid approach on clustering solutions obtained using fuzzy rough granular self-organizing map (FRGSOM) as described in Chap. 3. As many genes belong to multiple biological categories, the FRGSOM will be able to provide better clustering solutions than related methods where overlapping class boundaries exist. Hence better gene order is expected with the combination of FRGSOM and FRAG_GALK.

The approach for integrating multiple data sources for gene function prediction in *BS* involves local search in predicting the weights of the data sources. In this regard, heuristic base soft computing tools like GAs, simulated annealing and antcolony optimization may provide a better solution in approximating the weights through global search. However, it is likely that they will take much more computation time than any local search procedure. Function of each gene is predicted from its top *K* neighbors using *BS* as it is a well studied approach for gene function prediction. Here, application of FRGSOM can provide good clustering solutions. However, the no. of clusters will be less than the neighbor based approach where it is equal to the no. of genes which in turn leads to the function prediction of hundreds of genes. A good solution may be to predict clusters using both the methods separately on *BS* and then select the functionally enriched clusters.

6.6 Conclusion

Two methods involving integration of gene ordering with partitive clustering and integration of multiple biological data sources are described in this chapter. Their utility in finding useful groups of genes, grouping functionally correlated genes within clusters, and their function prediction are demonstrated on various data sets. The first approach not only determines unique clusters, but also preserves the biologically meaningful relationships/order among the genes within clusters. The second one reveals relationship between genes as well as their functions which could not be revealed using any single data source; thus pointing out the importance of heterogeneous data source integration.

In FRAG_GALK, parallel searching (with large population in genetic algorithm) for optimal gene order in gene clusters (closely related genes) is performed. While this results in reduced searching time for FRAG_GALK as compared to Concorde_LP and B-Joseph's method, in terms of biological score, FRAG_GALK is comparable with Concorde_LP and B-Joseph's method and sometimes superior to B-Joseph's method. It is evident from the experimental results that the combination of partitive clustering and FRAG_GALK is a promising tool for microarray gene expression analysis.

In *BS* a framework for data source integration, through functional annotation based weighting, is developed to predict gene function for yeast. Five data sources, namely, phenotypic profiles, gene expression data, KEGG profiles, protein protein interaction, and protein sequence similarity through transitive homologues are used. Functional prediction of 12 unclassified yeast genes confirmed the potential value of *BS*. The flexibility of *BS* also allows for easy inclusion of other data sources. The *BS* framework can also be extended to a larger test bed by including similarities arising from gene-fusion and gene-order conservation-based methods. As mentioned in Sect. 6.1, granular computing is not utilized in the methods described in this chapter, but it is a highly relevant framework in the context of gene expression analysis involving clustering techniques and prediction tasks.

References

1. Altschul, S.F., Madden, T.L., Schffer, A.A., Zhang, J., Zhang, Z., Miller, W., Lipman, D.J.: Gapped blast and PSI-blast: a new generation of protein database search programs. Nucleic Acids Res. **25**(17), 3389–3402 (1997)
2. Applegate, D., Bixby, R., Chvatal, V., Cook, W.: Concorde package. www.tsp.gatech.edu/ concorde/downloads/codes/src/co031219.tgz (2003)
3. Ashburner, M., Ball, C.A., Blake, J.A., Botstein, D., Butler, H., Cherry, J.M., Davis, A.P., Dolinski, K., Dwight, S.S., Eppig, J.T., Harris, M.A., Hill, D.P., Issel-Tarver, L., Kasarskis, A., Lewis, S., Matese, J.C., Richardson, J.E., Ringwald, M., Rubin, G.M., Sherlock, G.: Gene ontology: tool for the unification of biology, the gene ontology consortium. Nat. Genet. **25**(1), 25–29 (2000)

4. Bairoch, A., Apweiler, R., Wu, C.H., Barker, W.C., Boeckmann, B., Ferro, S., Gasteiger, E., Huang, H., Lopez, R., Magrane, M., Martin, M.J., Natale, D.A., O'Donovan, C., Redaschi, N., Yeh, L.S.: The universal protein resource (UniProt). Nucleic Acids Res. **33**, 154–159 (2005)

5. Bar-Joseph, Z., Gifford, D.K., Jaakkola, T.S.: Fast optimal leaf ordering for hierarchical clustering. Bioinformatics **17**, 22–29 (2001)

6. Barker, W.C., Garavelli, J.S., Huang, H., McGarvey, P.B., Orcutt, B.C., Srinivasarao, G.Y., Xiao, C., Yeh, L.-S.L., Ledley, R.S., Janda, J.F., Pfeiffer, F., Mewes, H.W., Tsugita, A., Wu, C.: The protein information resource (PIR). Nucleic Acids Res. **28**(1), 41–44 (2000)

7. Ben-Dor, A., Shamir, R., Yakhin, Z.: Clustering gene expression patterns. J. Comput. Biol. **6**, 281–297 (1999)

8. Biedl, T., Brejova, B., Demaine, E.D., Hamel, A.M., Vinar, T.: Optimal arrangement of leaves in the tree representing hierarchical clustering of gene expression data. Department of Computer Science, University of Waterloo (2001)

9. Brown, J.A., Sherlock, G., Myers, C.L., Burrows, N.M., Deng, C., Wu, H.I., McCann, K.E., Troyanskaya, O.G., Brown, J.M.: Global analysis of gene function in yeast by quantitative phenotypic profiling. Mol. Syst. Biol. **2**(1), 1–9 (2006)

10. Cleveland, W.S., Devlin, S.J.: Locally weighted regression: an approach to regression analysis by local fitting. J. Am. Stat. Assoc. **83**, 596–610 (1988)

11. Cotta, C., Mendes, A., Garcia, V., Franca, P., Moscato, P.: Applying Memetic Algorithms to the Analysis of Microarray Data. In: Raidl, G., Cagnoni, S., Cardalda, J.J.R., Corne, D.W., Gottlieb, J., Guillot, A., Hart, E., Johnson, C.G., Marchiori, E., Meyer, J.A., Middendorf, M. (eds.) Applications of Evolutionary Computing. Lecture Notes in Computer Science, pp. 22–32. Essex (2003)

12. Delisi, C., Yanai, I.: The society of genes: networks of functional links between genes from comparative genomics. Genome Biol. **3**(11), 1–64 (2002)

13. Dwight, S.S., Harris, M.A., Dolinski, K., Ball, C.A., Binkley, G., Christie, K.R., Fisk, D.G., Issel-Tarver, L., Schroeder, M., Sherlock, G., Sethuraman, A., Weng, S., Botstein, D., Cherrya, J.M.: Saccharomyces genome database (SGD) provides secondary gene annotation using the gene ontology (GO). Nucleic Acids Res. **30**(1), 69–72 (2002)

14. Eisen, M.B., Spellman, P.T., Brown, P.O., Botstein, D.: Cluster analysis and display of genome-wide expression patterns. Proc. Natl. Acad. Sci. USA **95**, 14,863–14,867 (1998)

15. Enright, A.J., Iliopoulos, I., Kyrpides, N.C., Ouzounis, C.A.: Protein interaction maps for complete genomes based on gene fusion events. Nature **402**, 86–90 (1999)

16. Gillespie, D., Spiegelman, S.: A quantitative assay for DNA-RNA hybrids with dna immobilized on a membrane. J. Mol. Biol. **12**(3), 829–842 (1965)

17. Hartwell, L.H., Hopfield, J.J., Leibler, S., Murray, A.W.: From molecular to modular cell biology. Nature **402**, C47–C52 (1999)

18. Hellem, B.T., Dysvik, B., Jonassen, I.: LSimpute: Accurate estimation of missing values in microarray data with least squares methods. Nucleic Acids Res. **32**(3), e34 (2004)

19. Herwig, R., Poustka, A.J., Muller, C., Bull, C., Lehrach, H., O'Brien, J.: Large-scale clustering of cDNA-fingerprinting data. J. Genome Res. **9**, 1093–1105 (1999)

20. Huynen, M.A., Bork, P.: Measuring genome evolution. Proc. Natl. Acad. Sci. USA **95**, 5849–5856 (1998)

21. Jenner, R.G., Alba, M.M., Boshoff, C., Kellam, P.: Kaposi's sarcoma-associated herpesvirus latent and lytic gene expression as revealed by dna arrays. J. Virol. **75**(2), 891–902 (2001)

22. Johnson, D.S., McGeoch, L.A.: The Traveling Salesman Problem: A Case Study in Local Optimization: Local Search in Combinatorial Optimization. Wiley, New York (1996)

23. Kanehisa, M., Goto, S., Hattori, M., Aoki-Kinoshita, K.F., Itoh, M., Kawashima, S., Katayama, T., Araki, M., Hirakawa, M.: From genomics to chemical genomics: new developments in KEGG. Nucleic Acids Res. **34**, D354–D357 (2006)

24. Kawasaki, S., Borchert, C., Deyholos, M., Wang, H., Brazille, S., Kawai, K., Galbraith, D., Bohnert, H.J.: Gene expression profiles during the initial phase of salt stress in rice. Plant Cell **13**(4), 889–906 (2001)

25. Krause, E.F.: Taxicab Geometry: An Adventure in Non-Euclidean Geometry. Dover, New York (1986)
26. Lee, I., Date, S.V., Adai, A.T., Marcotte, E.M.: A probabilistic functionalnetwork of yeast genes. Science **306**, 1555–1558 (2004)
27. Lee, I., Narayanaswamy, R., Marcotte, E.M.: Bioinformatic prediction of yeast gene function. In: Stansfield, I. (ed.) Yeast Gene Analysis. Elsevier Press, Amsterdam (2006)
28. Ma, Q., Chirn, G.W., Cai, R., Szustakowski, J.D., Nirmala, N.: Clustering protein sequences with a novel metric transformed from sequence similarity scores and sequence alignments with neural networks. BMC Bioinform. **6**(242) (2005)
29. Marcotte, E.M., Pellegrini, M., Ng, H.L., Rice, D.W., Yeates, T.O., Eisenberg, D.: Detecting protein function and protein-protein interactions from genome sequences. Science **285**, 751–753 (1999)
30. Marcotte, E.M., Pellegrini, M., Thompson, M.J., Yeates, T.O., Eisenberg, D.: A combined algorithm for genome-wide prediction of protein function. Nature **402**, 83–86 (1999)
31. Mering, C.V., Krause, R., Snel, B., Cornell, M., Oliver, S.G., Fields, S., Bork, P.: Comparative assessment of large-scale data sets of protein-protein interactions. Nature **417**, 399–403 (2002)
32. Mewes, H.W., Frishman, D., Gldener, U., Mannhaupt, G., Mayer, K., Mokrejs, M., Morgenstern, B., Mnsterktter, M., Rudd, S., Weil, B.: MIPS: a database for genomes and protein sequences. Nucleic Acids Res. **30**(1), 31–34 (2002)
33. Munich information center for protein sequences. http://mips.gsf.de/ (2008)
34. Pal, S.K., Bandyopadhyay, S., Ray, S.S.: Evolutionary computation in bioinformatics: a review. IEEE Trans. Syst. Man Cybern. Part C **36**(5), 601–615 (2006)
35. Park, J., Karplus, K., Barrett, C., Hughey, R., Haussler, D., Hubbard, T., Chothia, C.: Sequence comparisons using multiple sequences detect three times as many remote homologues as pairwise methods. J. Mol. Biol. **284**, 1201–1210 (1998)
36. Pawson, T., Nash, P.: Assembly of cell regulatory systems through protein interaction domains. Science **300**, 445–452 (2003)
37. Pellegrini, M., Marcotte, E.M., Thompson, M.J., Eisenberg, D., Yeates, T.O.: Assigning protein functions by comparative genome analysis: protein phylogenetic profiles. Proc. Natl. Aca. Sci. USA **96**, 4285–4288 (1999)
38. Pipenbacher, P., Schliep, A., Schneckener, S., Schonhuth, A., Schomburg, D., Schrader, R.: Proclust: improved clustering of protein sequences with an extended graph-based approach. Bioinformatics **18**(2), S182–S191 (2002)
39. Ray, S.S., Bandyopadhyay, S., Pal, S.K.: Gene ordering in partitive clustering using microarray expressions. J. Biosci. **32**(5), 1019–1025 (2007)
40. Ray, S.S., Bandyopadhyay, S., Pal, S.K.: Genetic operators for combinatorial optimization in TSP and microarray gene ordering. Appl. Intell. **26**(3), 183–195 (2007)
41. Ray, S.S., Bandyopadhyay, S., Pal, S.K.: Combining multi-source information through functional annotation based weighting: gene function prediction in yeast. IEEE Trans. Biomed. Eng. **56**(2), 229–236 (2009)
42. Reguly, T., Breitkreutz, A., Boucher, L., Breitkreutz, B.J., Hon, G.C., Myers, C.L., Parsons, A., Friesen, H., Oughtred, R., Tong, A., Stark, C., Ho, Y., Botstein, D., Andrews, B., Boone, C., Troyanskya, O.G., Ideker, T., Dolinski, K., Batada, N.N., Tyers, M.: Comprehensive curation and analysis of global interaction networks in Saccharomyces cerevisiae. J. Biol. **5**(4), 1–28 (2006)
43. Salwinski, L., Miller, C.S., Smith, A.J., Pettit, F.K., Bowie, J.U., Eisenberg, D.: The database of interacting proteins. Neuclic Acids Res. **32**, 449–451 (2004)
44. Sharan, R., Maron-Katz, A., Shamir, R.: Click and expander: a system for clustering and visualizing gene expression data. Bioinformatics **19**(14), 1787–1799 (2003)
45. Sherlock, G., Hernandez-Boussard, T., Kasarskis, A., Binkley, G., Matese, J.C., Dwight, S.S., Kaloper, M., Weng, S., Jin, H., Ball, C.A., Eisen, M.B., Spellman, P.T., Brown, P.O., Botstein, D., Cherry, J.M.: The stanford microarray database. Nucleic Acids Res. **29**(1), 152–155 (2001)
46. Southern, E.M.: Detection of specific sequences among DNA fragments separated by gel electrophoresis. J. Mol. Biol. **98**(3), 503–507 (1975)

47. Spellman, P.T., Sherlock, G., Zhang, M.Q., Iyer, V.R., Anders, K., Eisen, M.B., Brown, P.O., Botstein, D., Futcher, B.: Comprehensive identification of cell cycle-regulated genes of the yeast saccharomyces cerevisia by microarray hybridization. Mol. Biol. Cell **9**, 3273–3297 (1998)
48. Spirin, V., Mirny, L.A.: Protein complexes and functional modules in molecular networks. Proc. Natl. Acad. Sci. USA **100**(21), 12,123–12,128 (2003)
49. Stark, C., Breitkreutz, B., Reguly, T., Boucher, L., Breitkreutz, A., Tyers, M.: BioGRID: a general repository for interaction datasets. Nucleic Acids Res. **34**, D535–D539 (2006)
50. Tamayo, P., Slonim, D., Mesirov, J., Zhu, Q., Kitareewan, S., Dmitrovsky, E., Lander, E.S., Golub, T.R.: Interpreting patterns of gene expression with self-organizing maps: methods and application to hematopoietic differentiation. Proc. Natl. Acad. Sci. USA **96**, 2907–2912 (1999)
51. Troyanskaya, O.G., Dolinski, K., Owen, A.B., Altman, R.B., Botstein, D.: A Bayesian framework for combining heterogeneous data sources for gene function prediction (in Saccharomyces cerevisiae). Proc. Natl. Acad. Sci. USA **100**(14), 8348–8353 (2003)
52. Tsai, H.K., Yang, J.M., Tsai, Y.F., Kao, C.Y.: An evolutionary approach for gene expression patterns. IEEE Trans. Inf Technol. Biomed. **8**(2), 69–78 (2004)
53. Venet, D.: MatArray: a Matlab toolbox for microarray data. Bioinformatics **19**(5), 659–660 (2003)
54. Website. http://rana.lbl.gov/eisendata.htm (2008)
55. Wu, W., Xing, E.P., Myers, C., Mian, I.S., Bissell, M.J.: Evaluation of normalization methods for cdna microarray data by K-NN classification. BMC Bioinform. **6**(191), 1–21 (2005)
56. Xie, H., Wasserman, A., Levine, Z., Novik, A., Grebinskiy, V., Shoshan, A., Mintz, L.: Large-scale protein annotation through gene ontology. Genome Res. **12**, 785–794 (2002)
57. Yang, Y.H., Dudoit, S., Luu, P., Speed, T.P.: Normalization for cdna microarray data: a robust composite method addressing single and multiple slide systematic variation. Nucleic Acids Res. **30**(4), e15 (2002)

Chapter 7
RNA Secondary Structure Prediction: Soft Computing Perspective

7.1 Introduction

The availability of huge amount of biological data has opened a new direction in genomic analysis and structural prediction of deoxyribonucleic acid (DNA), ribonucleic acid (RNA) and proteins in recent years. The challenge is to find an efficient way to use these rich trove of evidence to infer functional, biological and structural properties. In Chap. 6 various types of biological datasets such as DNA microarray gene expression, protein sequence and phenotypic profile are analyzed. Many of these data, led to an absolute demand for specialized tools to view, analyze and predict the biological significance of the data. However, another important data namely RNA sequence is not used for analysis by any method described in this book so far. This chapter deals with the problem of RNA secondary structure prediction which is described in [59]. The problem gained significant attention of the researchers in the last few decades as it is one of the key issues in understanding the genetic diseases and creating new drugs. It also helps the biologists to understand the role of the molecule in the cell [40, 62, 72, 78]. As mentioned in Sect. 1.8, several deterministic algorithms and soft computing based techniques, such as genetic algorithms, artificial neural networks, and fuzzy logic, have been developed to determine the structure from a known RNA sequence but rough sets and granular computing are still not explored for this purpose. Soft computing is mainly utilized to get approximate solutions for RNA sequences by considering the issues related with kinetic effects, cotranscriptional folding and estimation of certain energy parameters [59].

There are different types of RNA(s), e.g., transfer RNA (tRNA), messenger RNA (mRNA), viral RNA, ribozomal RNA, signal recognition particle RNA (SRP RNA). In a cell, the role of different RNAs varies widely from each other. For example, RNA is used as a genetic material, instead of DNA, by some viruses, and messenger RNA (mRNA) is used by all organisms in proteins synthesis. Small nuclear RNAs (snRNA) are those eukaryotic RNAs which are involved in processing hnRNAs. Some RNAs are responsible for controlling gene expression, or sensing and communicating responses to cellular signals.

© Springer International Publishing AG 2017
S.K. Pal et al., *Granular Neural Networks, Pattern Recognition and Bioinformatics*, Studies in Computational Intelligence 712, DOI 10.1007/978-3-319-57115-7_7

The RNA secondary structure prediction problem is a critical one in molecular biology [59]. Secondary structure as well as tertiary structure can be determined by X-ray crystallography [4, 31] and Nuclear Magnetic Resonance (NMR) spectroscopy [17]. Techniques involving small-angle X-ray solution scattering (SAXS) [56], hydroxyl radical probing [29, 74], in-line probing [60], and modification of bases by selective 2′-hydroxyl acylation analyzed by primer extension (SHAPE) [41], dimethyl sulfate (DMS) [70], l-cyclohexyl-3-[2-morpholinoethl] carbodiimide metho-p-toluene sulfonate (CMCT) [2], 1,1-Dihydroxy-3-ethoxy-2-butanone (known as kethoxal) [50], nucleases [9, 28], diethyl pyrocarbonate [68] and ethylnitrosourea [68] are also successfully used to determine the secondary structure. In general, these processes are difficult, slow, and expensive. That is why developing computational methods to predict the secondary structure of RNA is very necessary. Data analysis tools used for prediction of RNA structure are mainly based on dynamic programming [1, 26, 52, 80, 94–96]. The role of soft computing tools, a collection of flexible information processing techniques [90], gained significance with the need to generate approximate and good solutions by considering the kinetic effects in RNA folding [3, 5, 22], cotranscriptional folding [22, 46, 61, 88], pseudoknots [67] and deficiencies in energy parameters [42]. In this chapter, we describe the role of various soft computing tools, like genetic algorithms (GAs), artificial neural networks (ANNs) and Fuzzy logic (FL), for formulating different methodologies as explained in [59]. While, GAs are adaptive, and robust search processes, producing near optimal solutions, and have a large amount of implicit parallelism in creating a number of solutions, artificial neural networks are machinery for adaptation & curve fitting and fuzzy logic deals with the concept of partial truth, approximate reasoning and partial set membership for uncertainty handling. GAs are adaptive in the search process as they identify the good and bad solutions in the search space according to the fitness of the chromosomes, select a set of good solutions probabilistically in every iteration, and recombine partial solutions (through crossover operation) from the best chromosomes to form chromosomes of potentially higher fitness. For a neural network, curve-fitting refers to the modeling of an approximate input output relationship, through supervised learning. Once the relation has been modeled, to the necessary accuracy, it can be used for structure prediction, with an input-output characteristic approximately equal to the relationship existing in the training set. For example, the curve fitting can be accomplished by training the neural network with vertex identification results of a tree like RNA structure [34].

This chapter presents the research works described in [59] for predicting RNA secondary structure using different components of soft computing, and enable the researchers in both, biology and machine learning, to understand the relevant problems and issues of each other for furtherance of bioinformatics research. First, we describe the basic concepts in RNA structure prediction in Sect. 7.2. The dynamic programming based methods for prediction of RNA structure are explained in Sect. 7.3. Sections 7.4 and 7.5 deal with the role of soft computing methods, like genetic algorithms (GAs), artificial neural networks (ANNs), and fuzzy logic (FL) in RNA secondary structure prediction problem. In Sect. 7.6, the relevance of metaheuristics like simulated annealing (SA) and particle swarm optimization (PSO) in RNA secondary structure prediction is described. Two machine learning techniques, k-nearest neighbor classifier and support vector machines (SVMs), which are widely used for RNA structure prediction are discussed briefly in Sect. 7.7. The performances of the aforesaid methods in predicting structures of several RNA sequences are compared in Sect. 7.8. Major challenges to bioinformatics community and soft computing tools, developed under neuro fuzzy framework, for solving the related prediction problems are discussed in Sect. 7.9. Finally, some conclusions and future research directions are presented in Sect. 7.10.

7.2 Basic Concepts in RNA

Here, we introduce the biological concepts required to understand the RNA structure [59], and then we describe the various secondary structural elements. The effect of ions, temperature and proteins on RNA is also provided.

7.2.1 Biological Basics

An RNA molecule represents a long chain of monomers called nucleotides, and each nucleotide consists of a base (any one of adenine (A), cytosine (C), guanine (G) and uracil (U)), a phosphate group and a sugar group [40]. The specific sequence of the bases along the chain is called primary structure. The structure is usually defined over the alphabets 'A', 'C', 'G', and 'U'. Through the formation of hydrogen bonds the two groups of complementary bases, A-U and C-G, form stable base pairs and are known as the Watson-Crick base pairs [36, 40, 81]. While, the A-U pairs form two hydrogen bonds, the C-G pairs form three hydrogen bonds and tend to be more stable than the A-U pairs. Other bases also sometimes pair, especially the G-U pair. The G-U pairs are known as wobble base pairs and form two hydrogen bonds [8, 78]. Note that, different types (depending on strand orientation) of U-U, U-C, G-A, A-A and A-C base pairs [38] and base triples like GAC, GGC, etc. are also found in RNA structures [14]. A base pair provides a groove for insertion of an unpaired nucleoside to form a triple. The base-pair structure is referred to as the secondary structure of RNA [27, 52, 71, 76]. Generally, the secondary structure is determined in terms of substructures by observing each base is either paired or not and structure formation for RNA molecules is dependent on the Watson-Crick base pairs, wobble base pairs and stacking of adjacent base pairs.

7.2.2 Secondary Structural Elements in RNA

Seven recognized secondary structural elements exist in RNA and these are (1) Hairpin loop, (2) Bulge loop, (3) Internal loop, (4) Multi-branched loop, (5) Single-

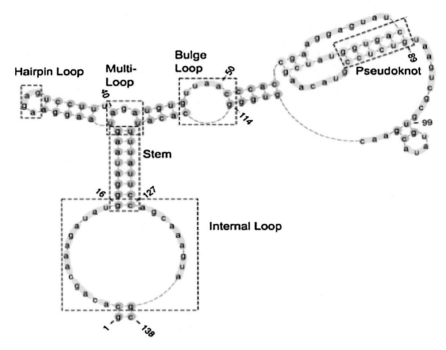

Fig. 7.1 Different types of secondary substructures in RNA

stranded regions, (6) Helix, and (7) Pseudoknots. A common sub-substructure also
exists in many parts of RNA, as well as in hairpin loop, helices, and pseudoknots,
called stem. When a stem is a part of a helix we can refer it as a helical stem.
A schematic view of various structural elements, within dotted boxes, is shown in
Fig. 7.1. Stems also help to identify the start and end of most substructures. Four-
letter alphabet is used to represent an RNA sequence, which is the primary structure
of RNA. Let $S = s_1, s_2, \ldots, s_n$ be an RNA sequence, where, base $s_i \in \{A, U, C, G\}$
and $1 \le i \le n$. The subsequence $s_{(i,j)} = s_i, s_i + 1, \ldots, s_j$ is a segment of S where,
$1 \le i \le j \le n$. If s_i & $s_j \in \{A\text{-}U, C\text{-}G, U\text{-}G\}$, then s_i and s_j may constitute a base
pair (i, j). Each base can at most take part in one base pair. Now, we describe, in
brief, different substructures within RNA secondary structure.

- Single-stranded region: Let r be a sequence of bases in an RNA sequence. If all
  the bases in r are not paired with any other bases in the RNA structure, then we
  say r is a single-stranded region.
- Stem: Contiguous stacked base pairs are called stems [83]. In this structure, base
  pairs are generally stacked onto other base pairs. If base pairs (i, j) and $(i + 1,$
  $j - 1) \in S$ then, base pairs (i, j) and $(i + 1, j - 1)$ constitute the stack $(i, i + 1 :$
  $j - 1, j)$, and $m(\ge 2)$ consecutive stacks form the stem $(i, i + m : j - m, j)$ with
  the length of $m + 1$.
- Loop: Single stranded subsequence bounded by base pairs is called a loop [27]

- Hairpin loop: A loop at the end of a stem is called a hairpin loop. If $(i, j) \epsilon$ S but none of the elements, $i + 1 \ldots j - 1$, are paired with any other base, then the cycle is called a hairpin loop. Many molecular biologists use "hairpin" to refer to a stem with a loop of size 0 or 1 at the end, i.e., a stem with virtually no loop.
- Bulge loop: Single stranded bases occurring within only one side of a stem are called a bulge loop. If substructure contains base pairs (i, j) and $(i + 1, q)$, and there are some unpaired elements between q and j, then these unpaired elements form a bulge loop.
- Internal loop: In an internal loop there are single stranded bases interrupting both sides of a stem. If $i + 1 < p < q < j - 1$ and substructure contains base pairs (i, j) and (p, q), but the elements between i and p are unpaired and the elements between q and j are also unpaired, then the two unpaired regions constitute an internal loop.
- Multi-branched loop: The loop from which three or more stems radiate is called multi-branched loop [37]. If substructure contains two or more base pairs like (i, j), (p, q), and (r, s), and the indices of none of the pairs lie within each other, then a multi-branched loop is formed.
- Helix: In general, stems are considered to be helices [83], and they provide stability in the secondary structure. Single stranded RNA folds back itself, forming helical areas interspersed with unpaired, single-stranded areas.

  Since the generation of a helix terminates at the first mismatched base pair, other secondary structures are implicitly defined in the various bulges and loops which remain outside of the stacked pairs. Thus, the determination of helices alone is considered sufficient, in some investigations [73, 85, 86], to account for all other secondary structure elements.
- Pseudoknots: Pseudoknots occur when unpaired bases of one substructure (e.g., the loop part in a hairpin loop) bind to unpaired bases of another substructure to form a stem [83]. If the resulting stem, formed from this type of bonding, stacks upon an existing stem, then the new formation is called coaxial stacking between two stems with a quasicontinuous helix. This structure also forms a pseudoknot and helps to stabilize the RNA structure. Even in pseudoknots, the major driving force of structure formation is Watson-Crick base pairs, Wobble base pairs, and stacking of adjacent base pairs.

In RNA the primary sequence determines the secondary structure which, in turn, determines its tertiary folding [72]. The secondary structural elements interact between themselves through formation of hydrogen bonds and Van der Waals interactions and fold in a three dimensional space to form the tertiary structure, a biologically active conformation. The folding process is facilitated by the presence and increasing concentration of cations (like magnesium ions), lowering temperature, and presence of proteins, called RNA chaperones [13, 82]. The interactions can take place between two helices, two unpaired regions, or one unpaired region and a double-stranded helix [82].

Although, the tertiary structure is the level of organization relevant for biological function of structured RNA molecules and sometimes secondary structures are influenced by tertiary structures [33, 59], the interactions, responsible for RNA tertiary structure formation, are significantly weaker than those responsible for secondary structure formation [72]. Hence in recent investigations [20, 42, 44], for computational simplicity it is assumed that the influence of hydrogen bonds within tertiary structure, on hydrogen bonds within secondary structure, is negligible; consequentially, secondary structures can be predicted independently of tertiary structures. Current research is going on SHAPE (Selective 2'-hydroxyl acylation analyzed by primer extension) [41], a powerful biochemical method for RNA structure probing, that uses reagents to modify the backbone in structurally flexible regions of RNA, and sequencing based characterization of RNA structure [21].

7.2.3   Example

Gibbs free energy ($\triangle G$) is generally used for calculating the total energy of different RNA structures (obtained from the same sequence) and the structure with minimum energy is accepted. This energy is a thermodynamic potential that measures the capacity of a system to do non-mechanical work. It is also the chemical potential that is minimized when a system reaches equilibrium at constant pressure and temperature. Hence, it is widely used to calculate fitness of the RNA secondary structure, obtained using various prediction algorithms [15]. It is represented as a function of enthalpy ($\triangle H$), temperature T, and entropy ($\triangle S$):

$$\triangle G = \triangle H - T\triangle S. \tag{7.1}$$

Enthalpy is a measure of the total energy of a thermodynamic system. The total energy is the sum of the internal energies. It is also the energy required to create a system, and the amount of energy required by the system for displacing its environment and establishing its volume and pressure. Temperature is a measure of the average kinetic energy in a system and entropy is a measure of how much of the energy of a system is potentially available to do work and how much of it will potentially manifest as heat. So, change in entropy can be used as a quantitative measure of the relative disorder of a system. Differences in Gibbs free energy ($\triangle G$), in a reaction or a conformation change, provide information on the process spontaneity. A positive free energy difference indicates that the reaction is in favor of the reactants and the reaction will go on spontaneously. A negative free energy difference in a reaction favors the products. A zero value of the free energy indicates that neither the reactants nor the products are favored.

Now, we will show how free energy can be calculated for a helix formation in an RNA secondary structure, using free energy thermodynamic parameters [43, 59]. Let $\binom{5'GAUC3'}{3'CUAG5'}$ represent a helix within an RNA structure. The free energy change for this helix formation can be computed as:

$$\Delta G(pred) = 2\Delta G \begin{pmatrix} 5'GA3' \\ 3'CU5' \end{pmatrix} + \Delta G \begin{pmatrix} 5'AU3' \\ 3'UA5' \end{pmatrix}$$

$$+ \Delta G_{init} + \Delta G_{AU\ end\ penalty} (per\ AU end) + \Delta G_{sym}. \tag{7.2}$$

Using the values, downloaded from TURNER LAB, as mentioned in [43, 75], the free energy becomes

$$\Delta G(pred) = 2(-2.4) + (-1.1) + (4.09) + 0 + 0.43$$

$$= -1.38\,\text{kcal/mol} \tag{7.3}$$

In this example, the nearest neighbor terms are generated by considering a sliding window with two adjacent base pairs at the duplex RNA structure, and these results in three terms $\begin{pmatrix} 5'GA3' \\ 3'CU5' \end{pmatrix}$, $\begin{pmatrix} 5'AU3' \\ 3'UA5' \end{pmatrix}$, and $\begin{pmatrix} 5'UC3' \\ 3'AG5' \end{pmatrix}$. The last term $\begin{pmatrix} 5'UC3' \\ 3'AG5' \end{pmatrix}$ is similar to $\begin{pmatrix} 5'GA3' \\ 3'CU5' \end{pmatrix}$, as one can be obtained by rotating the another by 180°. There is a loss of entropy during initial pairing between the first two bases. This is accounted by a constant initiation term ΔG_{init}. The term $\Delta G_{AU\ end\ penalty} (per\ AU end) = 0$ as there are no AU base pair at the end for this helix. Furthermore, if there exists a GG mismatch (start of a loop) after the last AU pair then a bonus term is initiated. The last term ΔG_{sym} corrects for twofold rotational symmetry, resulting from self complementary strand. If the two strands in the helix are not complementary then the last term is not considered for energy calculation. Moreover, if there is any GU base pair at the end of a helix then the $\Delta G_{GU\ end\ penalty}$ should be taken into account. The free energy change for a hairpin loop with 4 or more unpaired nucleotides can be computed as:

$$\Delta G(hairpin) = \Delta G_{init}(n) + \Delta G_{terminal\ mismatch}$$

$$+ \Delta G_{GA\ or\ UU\ first\ mismatch} + \Delta G_{GG\ first\ mismatch} \tag{7.4}$$

$$+ \Delta G_{special\ GU\ closure} + \Delta G_{penalty}(all C loops),$$

where, n is the number of nucleotides in the loop part only, 'terminal mismatch' parameter is the first mismatch stacking on the terminal base pair of a helix and not initiated for hairpin loop with 3 unpaired nucleotides, 'GA or UU first mismatch' and 'GG first mismatch' parameters are bonus terms, 'special GU closure' is applicable to hairpins where a GU closing pair (not UG) is preceded by two Gs, and the last term assigns a penalty for a loop with all C nucleotides. A tutorial for free energy calculation in various types of base-pairs, helix, terminal mismatches, loops and coaxial stacking is available at http://rna.urmc.rochester.edu/NNDB/turner04/index. html.

7.3 Dynamic Programming for RNA Structure Prediction

As dynamic programming algorithm (DPA) is first used to predict RNA secondary structure and is also used to predict substructures in some soft computing based methodologies, it will be prudent to have an idea of this algorithm and how the research started in predicting RNA secondary structures [59]. Dynamic programming (DP) [1, 52, 79, 80, 94, 96] is a computational way to solve optimization problems which can be divided into subproblems. At first, the optimal solution for each independent subproblem is calculated and then the solutions for overlapping subproblems are calculated by a recursive algorithm, repeatedly [79]. The optimal solutions for the subproblems are then preserved. The most probable secondary structure is predicted by calculating the sum of free energies, available from each optimal substructure, for all possible combinations of substructures and the combination with minimum free energy is accepted.

One of the first attempts to predict RNA secondary structure, using DPA and by maximizing the number of base-pairs using a simple nearest-neighbor energy model, is presented in [79]. Nussinov et al. [52] extended the method further. Zuker and Stiegler developed a popular dynamic programming based algorithm, mfold [95, 96], for finding the minimum free energy (MFE) pseudoknot-free secondary structure. In [79], an iterative definition of all secondary structures is first formulated and then functioned by minimizing the "distance" between segments of an RNA sequence, where, "distance" is measured in terms of free energy. It is also assumed that the formation of a given base pair is independent of all other base pairs. The initial steps are based on the research work of Needleman and Wunch [48]. Then the base pairing matrix, $p = (p_{i,j})$, for a given RNA sequence $s = s_1 s_2 \ldots s_n$ (and the reversed order sequence $s' = s_n s_{n-1} \ldots s_1$), is defined by $p_{i,j} = 1$ if s_i and s_j can form a bond, and $p_{i,j} = 0$ otherwise. When s_i and s_j are bonded, any bonding of $s_k (i < k < j)$ must be with points between i and j. The total number of structures, having $i + 1$ bonded pairs for a sequence $n + 1$ long, is given by a recursion relation. Let $N_{l,n}^i$ be the number of secondary structures, containing exactly i bonded pairs, formed on the subsequence $s_i s_{i+1} \ldots s_n$. Then

$$N_{l,n+1}^{i+1} = N_{l,n}^{i+1} + \sum_{j=l}^{n-m} \sum_{k=0}^{i} N_{l,j-1}^{k} N_{j+1,n}^{i-k} P_{j,n+1}, \qquad (7.5)$$

where, all hairpin loops have at least m bases. The equation follows from the fact that s_{n+1} is either bonded or not bonded. If s_{n+1} is not bonded, then there are $N_{l,n}^{i+1}$ structures of interest. Otherwise, $n + 1$ is bonded to some $j, l \leq j \leq n - m$, and if k bonds are formed in $s_l \ldots s_{j-1}$, then $i - k$ must be formed in $s_{j+1} \ldots s_n$. The definition of secondary structure implies that any combination of a k bonded structure with an $i - k$ bonded structure gives a secondary structure.

The mfold (multiple fold) webserver [95] uses the primary RNA sequence as input and predicts pseudoknot-free secondary structure with the minimized free energy (MFE) and some suboptimal structures. There are also options for users to choose the window for suboptimal structures, calculated in terms of percentage of the free energy of the minimum free energy structure, and to force some selected base pairs in the energy calculation process to consider some auxiliary information into account. The core algorithm uses the dynamic programming method and provides the energy dot plot matrices for the base pairs contained within the foldings. In general, the mfold server provides an interactive medium to the user to select an window for suboptimal structures, certain base pairs and number of solutions at the output.

Dynamic programming is also applied for calculating the equilibrium partition function for secondary structure, where the partition function is defined in terms of free energy, number of substructures and temperature [45]. The partition function is used to calculate the probabilities of various substructures in terms of base pairs. The effect of partition function with increase in temperature, in unfolding transition of RNA, is also studied. The method provides an ensemble of secondary structures.

For a given RNA sequence, the software package Sfold (statistical fold) [10] computes the partition function for the ensemble of all possible secondary structures and draws samples according to their Boltzmann equilibrium probability, to form the Boltzmann ensemble [12]. Different clusters are then produced from it and the centroid of the best cluster is chosen to provide the possible secondary structure [11].

The program RNAfold (RNA fold) from Vienna RNA package [25, 26] predicts a reliable secondary structure by checking the similarity between two structures, obtained using dynamic programming under MFE model and centroid of the best cluster in Boltzmann ensemble. The Vienna RNA package also provides tools for RNA comparison, structural alignments, prediction of RNA-RNA interactions and getting a clear perception about folding kinetics.

The KineFold [88] web server can predict RNA structures with pseudoknots by considering the co-transcriptional folding on time scale. The method is based on addition or removal of single helices and individual base-pair stacking/unstacking processes, as they are faster than nucleic acid folding and unfolding.

Though DPA traditionally yields optimal or suboptimal structures with minimum free energy, it shows limitations in considering the kinetic effects related to easily accessible states in RNA folding [3, 5, 22], states with high energy barrier [5, 63], and conditions concerning suitability of transition from one folding state to another. Cotranscriptional folding is another unaddressed issue in DPA, where, the partial RNA sequence starts folding before the entire sequence has been transcribed [22, 46, 61, 63, 88]. These have prompted researchers to use soft computing tools, which can handle kinetic effects in RNA folding, consider cotranscriptional folding, predict certain pseudoknots [67], and overcome some deficiencies in energy parameters [42] in a better fashion.

7.4 Relevance of Soft Computing in RNA Structure Prediction

Soft computing is a consortium of methodologies, that works synergistically and provides, in one form or another, flexible information processing capabilities for handling real life problems. Its aim is to exploit the tolerance for imprecision, uncertainty, approximate reasoning and partial truth in order to achieve tractability, robustness, low solution cost, and close resemblance with human like decision-making. The constituents of soft computing are mainly Fuzzy Logic (FL), Artificial Neural Networks (ANNs), Evolutionary Algorithms (EAs), and Rough Sets (RS) [54]. The present article mainly concerns with the application of Genetic Algorithms (GAs), ANNs, FL and some metaheuristics in RNA secondary structure prediction problem. The role of support vector machines (SVMs), which recently has gained attention of the researchers in pattern recognition (PR) and ANN is also discussed. At present, methods involving rough sets, recurrent neural networks, radial basis function neural network, genetic programming, evolutionary strategies, and evolutionary programming are not available in the literature, for RNA secondary structure prediction. Although, some of these methods are utilized for clustering or comparing the similarity of RNA secondary structures, generated from various prediction algorithms, they are not structure prediction algorithms. Hence, these methods are not considered here.

As mentioned in Sect. 7.3, DPA based methods try to find low energy stable structures by neglecting the issues related with kinetic energy barriers, cotranscriptional folding etc. Studying every possible structure for a sequence would solve the RNA folding problem, but it is not always feasible. However, one may handle this problem in soft computing paradigm where, GAs can be used to navigate in the landscapes of structures and can provide a set of possible solutions and ANNs can be trained with models of known RNA structures and can predict a structure if it resembles with a previously trained model. FL can be used in conjunction with DPAs, GAs or metahuristics to adjust various parameters in the prediction process. On the other hand, multiple crossover and mutation operations, for a predefined population size, can be run in parallel using GAs to approximately solve the problem. Note that, support vector machines, a machine learning method, are used with an alternative approach, where, RNA secondary structures can be predicted by comparative sequence analysis using functionally related sequences. However, soft computing based methods are not deterministic and provide a number of suboptimal solutions. Therefore, one has to choose a consensus structure from the given solutions. In this regard, the interactive web interface in mfold server [95], Stem Trace visualizer [30] and analyzing histogram peak energy values for iteration based methods may provide a direction to determine the consensus results. Although, the mfold server is designed to handle the dynamic programming results, the interface also allows an user to select base pairs, which can be chosen from the results of other methods.

Now, we discuss the characteristics of three soft computing methods, genetic algorithms [6, 19, 53], artificial neural network [23, 57], and fuzzy logic [89], that have been used in predicting the RNA secondary structure.

7.4.1 Characteristics of Different Soft Computing Technologies

The basic characteristics of different soft computing tools like genetic algorithms, artificial neural networks and fuzzy logic are described here.

7.4.1.1 Genetic Algorithms

Genetic Algorithms (GAs) [6, 19, 47] are adaptive heuristic search algorithms and premised on the evolutionary ideas of natural selection and genetics. The basic concept of GAs is to simulate processes in natural evolution that follow the principles of survival of the fittest. For optimization problems, GAs can provide robust, near optimal and fast solutions. They also have a large amount of implicit parallelism and provide a user defined number (population of chromosome) of alternative close approximate solutions. Therefore, for many of the real world problems, that involves finding optimal parameters and might prove difficult for traditional methods, the application of GAs appear to be an automatic choice. Since, GAs show outstanding performance in optimization, it can be used for classification and clustering problems. GAs are proved to be useful and efficient when

(a) the search space is highly complex and large to perform an exhaustive search, and
(b) conventional search methods cannot provide good solutions in a reasonable time.

7.4.1.2 Artificial Neural Networks

Artificial neural network (ANN) [23, 91] is a system composed of many simple processing elements (nodes) operating in parallel and is designed to emulate the biological neural network. ANNs are mainly used for function approximation, classification, prediction, feature extraction, and clustering. Depending on the task, the network can be supervised (SNN) or unsupervised (UNN). In SNN, the useful features within the data set are incorporated in the structure-activity relationship model of he network by training it with known useful features. This in turn, enables the network to detect correlations between the second and higher order patterns within data, and finds the application of SNN in biological systems, showing nonlinear behavior. Using unsupervised neural networks (e.g., Kohonen self-organizing maps), tasks like feature extraction and data clustering, can be performed without knowing the class information of the data points. The main advantage of UNN is embedded in its unsupervised learning where, no previous knowledge about the data is required. The main characteristics of ANNs are:

(a) it can easily adapt itself with the new patterns within data,
(b) encodes relation between the input and the output, however complicated, into network weights,
(c) tolerance to distorted patterns (ability to generalize),
(d) failure of components/nodes do not affect the performance of the system,
(e) it can achieve high computational speed by using its components in parallel way, and
(f) error can be minimized through learning from examples (if input is A then output is B).

7.4.1.3   Fuzzy Logic

Fuzzy logic (FL) deals with many-valued logic where, the reasoning is approximate rather than fixed and exact [89]. While, in traditional logic theory binary sets and crisp sets have two-valued logic, true or false, in fuzzy logic the concept of partial truth is incorporated which allows partial set membership, rather than crisp set membership. The variables or membership values in fuzzy logic may have any value that ranges in between 0 and 1. It can also be implemented using non-numeric linguistic variables, such as low, medium and high, where, the variables may be managed by specific functions. The key idea is to relate the output with the inputs using if-statements. For example, if two bases are A and U then the membership value for pairing is high (or 1.0) and if they are A and A then, the same may be low (or 0.3), depending on the application domain. Some important characteristics of fuzzy logic are:

(a) definite conclusions can be drawn from complex systems that generate vague, ambiguous, or imprecise information
(b) exact reasoning is viewed as a limiting case of approximate reasoning, and
(c) any logical system can be fuzzified.

7.5   RNA Secondary Structure Prediction Using Soft Computing

We now describe the existing techniques and methodologies developed for predicting RNA secondary structure, using different soft computing tools.

7.5.1 Genetic Algorithms

Prediction of RNA secondary structure, using GAs, is investigated in [3, 22, 63–66, 84–86]. While, some of the implementations use a binary representation for encoding the possible solutions (chromosomes in GAs) [3, 22], real coded representations, considering each substructure as separate integer, are also used as solutions in some investigations [85, 86]. In general, GAs generate conflicting stems (e.g., different stems sharing the same base pairs) which require removal of one of the conflicting stems by a repairing mechanism at later stage.

Van Batenburg et al. [3] developed a genetic algorithm to predict RNA secondary structure, based on RNA folding pathways and free energy minimization. The method first creates a list of stems and a population of several possible solutions, each represented by a stem-array. If a stem is present in a solution then the corresponding position in the stem-array is marked with 1, and otherwise 0. The method proceeds with standard selection, mutation and crossover operations in binary GAs. The kinetic effects of RNA folding is incorporated by initially restricting the GA to a small part of the RNA sequence and then gradually increasing the sequence length by 10% of the initial string with each iteration of the GA. This resulted in inclusion of the whole RNA sequence in 10 iterations and temporary stems which could be partially disrupted by another stem depending on the free energy values of the competing stems as well as on the loop that is formed when the stem is added. Finally, the algorithm is further improved by assigning different values for a 'mutation that deletes a stem' & a 'mutation that adds a stem' and squaring the fitness values of all stems to favor bigger stems. An improvement of this method is suggested in [22] where, the kinetic character of stem formation and disruption is formulated by using probabilities, depending on the energy contribution of the stem. The concept of cotranscriptional folding is also incorporated by controlling the sequence length increment with the rate of energy improvement, obtained from the current GA iteration as compared to the previous one.

A massively parallel genetic algorithm for RNA folding, based on free energy minimization and implemented in a computer with 16,384 processors & SIMD (single instruction/multiple data) architecture, is described in [65]. These processors are arranged in a two-dimensional mesh with toroidal wraparound. At first, a stem pool is created using GA by considering each stem as a 4-tuple $(start, stop, size, energy)$ where, b_i is the ith base in the RNA sequence, base $b_{start+i}$ pairs with base b_{stop-i} for $0 \leq i \leq size$, and 'energy' represents the stacking energy of the stem. Stems are only generated for a user given size or larger. Each processor then randomly selects a stem from the stem pool and goes on adding stems to complete a chromosome, which enables the GA to navigate in the landscape of structures. Finally, the secondary structure for each chromosome is represented by sorting these stems (with 4-tuples) w.r.t. the $start$ position and preserving it in a region table. For any two conflicting stems, the second stem is removed from the chromosome. Mutations are performed before the crossover operation by selecting a stem randomly from the stem pool and inserting it in a chromosome, and possible conflicts are handled in

the crossover process. Uniform crossover operations are then performed by selecting parents, within two to eight neighboring processors and itself, using a ranked
selection criterion. A stem is transferred from a parent to a child if there is no con-
flict with already existing stems in the child. It is also pointed out that, in case of
DPA, the selection of optimal substructures, emanating from a multibranch loop,
may contribute more unpaired bases than global suboptimal solutions provided by
GAs. An improved annealing based mutation operator for this method is described
in [66] where, the total number of mutations drops linearly with generations of GA
as the mutation probability descends hyperbolically with the size of the secondary
structures.

The concept of incorporating pseudostems, in the stem pool of GA, is introduced
in [63] to accommodate multiple folding events and collision of stems in an RNA
sequence. A pseudostem is a pair of two stems separated by an internal loop, across
which coaxial stacking of base pairs occurs. Pseudostems ensure that some of the
new structures will be fit enough to survive in the selection process of GA and
help it to explore through the structures which lies within the energy barrier for a
traditional GAs and dynamic programming algorithms. It is also shown that while,
GA with high population size (e.g., 128k) predicts the rod-like linear structure, a
lower population size (e.g., 4k) mostly predicts the metastable structures, and for
intermediate population sizes the ratio of rod-like structures to metastable structures
increases with population size. In a related investigation in [64], the accuracy of GA,
fitness of chromosomes and convergence time of GA, for various population sizes
are studied for different datasets.

Wiese et al. [85] formulated the structure prediction task as a permutation of possible helices, using real coded GAs. At first, all potential helices are generated from
a given RNA sequence by a helix generation algorithm [87], using a thermodynamic
model. Each helix is then indexed with an integer ranging from 0 to $n - 1$, where, n
is the total number of generated helices. Each chromosome of GA is encoded by a
permutation of these integers and provides a solution for RNA secondary structure.
In any chromosome, if two or more helices share some common base-pairs then for
any two conflicting helices, the second helix is removed by checking all helices in
a chromosome from left to right. Selection, crossover, and mutation operations are
then applied to the chromosomes in a elitist model framework. At the end of the
process, a set (population) of chromosomes with high scores, i.e., the chromosomes
with minimized free energies, are accepted as possible solutions. Precautions are
taken such that GAs not only optimize RNA structure in terms of minimum free
energy, but also ensure that predicted structures are chemically feasible ones, i.e.,
any structure should not contain helices that share common bases.

The investigation in [85], is similar to that in [86] where, a population of chromosomes evolves by selection, crossover, and mutation. The main difference between
[86] and the recent method [85] has been the use of better crossover and mutation
operators and incorporating state-of-the-art thermodynamic models to calculate the
free energies. In [85], experimental results are provided by comparing the predicted
structures with 19 known structures from four RNA classes.

In general, GAs provide a population of solutions as sub-optimal structures and also make it possible to investigate not only the minimum free energy structure but also other structures that may be closer to the natural fold. GAs seem suited to implementation for RNA structure prediction and also for estimating certain energy parameters.

7.5.2 Artificial Neural Networks

An RNA structure prediction method, involving graph-theoretic tree representation of RNA and training a 3 layered back propagation artificial neural network (ANN) with vertex identification results of the tree, is presented in [34]. In the tree, stems are represented as edges, hairpins as vertex of degree one, internal loops and bulges as vertices of degree two, and junctions as vertices of degree three or more, using the technique of Le et al. [35]. The method is based on the hypothesis that tree representation of a secondary RNA structure can be achieved by merging the tree representation of smaller structures, out of many combinatorial possibilities. These include structures of known RNA, RNA-like candidates and not RNA-like candidates. The vertex identification results for the ith tree, resulting from the tree merge operation, are represented by a vector

$$p^i = \langle p_1^i, p_2^i, p_3^i, p_4^i \rangle, \tag{7.6}$$

where, ith tree is one of the training samples. Components of p^i correspond to four input nodes of a 3-layered neural network with 24 nodes in the input layer and two nodes (say, y_1 and y_2) at the output layer. The neural network is trained to identify a tree as RNA or not. Here, $y_1 = 1$ and $y_2 = 0$ correspond to an RNA-like tree and $y_1 = 0$ & $y_2 = 1$ if the tree is not RNA-like. The initial weights of hidden and output layer nodes are randomly assigned to values close to 0 and the error function is considered as

$$E = \frac{1}{2} \sum_{i=1}^{n} \| y(p^i) - q^i \|^2, \tag{7.7}$$

where $y(p^i) = \langle y_1(p^i), y_2(p^i) \rangle$ is the output corresponding to an input vector p^i, n is the total number of training samples (trees), $q^i = \langle 1, 0 \rangle$ if the tree corresponds to known RNA or RNA-like structure and $q^i = \langle 0, 1 \rangle$ for not RNA-like structure. The weights are updated according to the standard procedure of back-propagation neural network and training continues until the error is close to 0. Finally, new tree structures are presented to the network for prediction task. The main advantage of this method lies in predicting new RNA structures, if they matches with known examples. A related investigation is available in [24].

In [39], the number of base pairs of RNA is maximized by using a Hopfield neural networks (HNN) and circular graph representation. This representation was first introduced by Nussinov et al. [51] where, the nucleotides are first aligned along the circumference of a circle graph and then the base pairs are represented by circular arcs that link paired bases. The number of neurons in HNN are considered to be the same as the number of base pairs, represented by the arcs of the circle. Each neuron is assigned the following binary function:

$$O_i = 1 \text{ if } I_i > 0, \text{ and } 0 \text{ otherwise,}$$

where, O_i and I_i are the output and input of the ith neuron, respectively. $O_i=1$ indicates that the ith arc and the corresponding base pair are not included in the circle graph and vice versa. The neurodynamical model of ith McCulloch-Pitts neuron is represented as:

$$dI_i/dt = A(\Sigma_j^n d_{ij}(1 - O_j)(distance(i))^{-1})(1 - O_j)p(i)^{-1}$$

$$- Bh(\Sigma_j^n d_{ij}(1 - O_j))O_i p(i), \tag{7.8}$$

where, $d_{xy}=1$ if xth arc and the yth arc intersect each other in the circle graph, 0 otherwise, $h(x)$ is 1 if $x = 0$, 0 otherwise, A & B are the transfer functions and $p(i)$ is the absolute value of free energy of the ith base pair. A cost function, termed energy, is also introduced where, a neurons's contribution to the energy is measured by the following equation:

$$\Delta E_k = E(a_k = 0) - E(a_k = 1) = (\Sigma_i a_i \omega_{ki}), \tag{7.9}$$

where, a_k is the activation level of the ith neuron, and ω_{ki} is the connection weight between the ith and jth neuron. The state (on/off) of a neuron, in the network, is determined by the network itself and that state is selected which lowers the network's energy.

In [93], class information of RNA in the initialization of Hopfield network is introduced. This resulted in improvement of experimental results with respect to the related investigation in [39].

7.5.3 Fuzzy Logic

A fuzzy dynamic programming algorithm [16] is used to determine the RNA secondary structure in [69]. At first, multiple up-triangular matrices, each having combination of all possible base pairs, are constructed to store various substructures. Each element (i, j) in the up-triangular matrices is represented by a 4×4 matrix, mentioning the membership values for 16 possible base pair interactions (AA, AC, $\cdots$ UU). Fuzzy sets are separately used to partition the rows and the columns of the up-triangular matrices and known distributions of single bases from homologous RNA structures are incorporated as prior knowledge. For a particular position

(i, j), the membership value is calculated by using the position specific membership information from all up-triangular matrices. The fuzzy dynamic programming algorithm then iteratively updates the position information in all matrices and expands the base pair structure to predict the optimal structure such that, the product of the membership values, over all positions, are maximized.

7.6 Meta-Heuristics in RNA Secondary Structure Prediction with

In this section we discuss the role of different metaheuristics like simulated annealing (SA) and particle swarm optimization (PSO), in RNA secondary structure prediction. Metaheuristics are closely related to Genetic Algorithm, one of the components of soft computing, in the sense that they are computational method that optimizes a problem by iteratively trying to improve an initial solution with respect to a given measure of quality.

7.6.1 Simulated Annealing

The use of SA [32] for predicting RNA secondary structures, using the free energy minimization approach, was first described by Schmitz and Steger [61]. The secondary structure is predicted by iterative formation and disruption of single base pairs through SA. Consequently, the energy changes are either changes in free energy or free activation energy. At the beginning, random formation and disruption of base pairs are allowed and the resulting unfavorable energy structures are subsequently suppressed by using a probabilistic selection process, based on Boltzmann factor. While, the favorable structures are always accepted, the probability of accepting the new structure, with energy (E_{new}) greater than the old one (E_{old}), is computed by

$$Probability[Accept] < e^{-(E_{new}-E_{old})/R\theta}, \tag{7.10}$$

where, temperature R is the Boltzmann's Constant and θ is the "distribution parameter", decreased gradually with each step of base pair selection method. The whole process is repeated for a predefined number of iterations or until θ achieves the desired value. The investigation also provides an idea of "sequential folding" during transcription by considering RNA polymerase chain elongation rates.

SARNA-Predict [73] employs a modified SA as its search engine, combines a novel mutation operator and uses a free energy minimization based approach. The method first creates initial solutions as a permutation of helices. New structures are then generated by mutating the existing ones and all new structures with reduced amount of free energy are accepted. The mutation is accomplished by multiple swap operations between two randomly chosen helix positions. The process maximizes the chance of generating a new structure, which may not be achieved, using a single swap operation, due to the repairing process performed after mutation to obtain a valid RNA structure. The number of mutations is chosen as the product of the percentage of the total number of available helices and the current annealing temperature and hence varies with time. New structures with increased energy are also accepted with some probability, determined by the Boltzmann distribution, to avoid local minima in the search space. The probability of accepting a new structure with increased energy is computed by

$$Probability[Accept] = e^{-(E_{new}-E_{old})/T} = e^{-\Delta Cost/T}, \qquad (7.11)$$

where, temperature T is the current temperature and E is the energy state. According to Eq. 7.11, the energy of a system at temperature T, is probabilistically distributed among all different energy states in thermal equilibrium condition. In general, the process mostly accepts a downward step and sometimes accepts an upward step.

7.6.2   Particle Swarm Optimization

A Set-based Particle Swarm Optimization algorithm (SetPSO), using mathematical sets to predict RNA structures with minimum free energy, is described in [49]. For a given RNA sequence, all possible stems are first generated to form a universal set U. The secondary structure is then represented as a permutation of stems. Each stem is represented as a particle and the permutation of the particles is the vector representation of the PSO. While, in generic PSO the position allocation of each particle (stem for RNA) and searching the best position can be performed by updating the position and velocity vectors, respectively, in SetPSO these are updated using two sets, instead of vectors. The first set is an open set which contains elements that should be removed (subtracted) from the current position set. The second set is formed by adding a random subset of current position and a random subset of U. The traditional addition and subtraction operators are replaced respectively by the union and minus operations of the set. The personal best position of a particle is tracked by the particle itself during the update process and the final position set, a subset of U, provides a potential solution. Although it is indicated in [49] that DP based mfold provides more accurate structures than SetPSO, according to citation information it gained attention of the researchers in predicting RNA structures using PSO.

7.7 Other Methods

In this section we discuss in brief about two machine learning techniques, k-nearest neighbor classifier and support vector machines (SVMs), those have gained attention of the researchers in RNA.

The classification process of a data point, using k-nearest neighbor algorithm (k-NN), is based on the majority voting among its k closest points in the feature space. Based on this classifier, the software package KnetFold (k-nearest neighbor classifiers based Folding) predicts a secondary structure from alignment of multiple sequences. First, it predicts if any two columns of the alignment correspond to a base pair, using a voting scheme from the outputs of a hierarchical network of k-nearest neighbor classifiers, and then it generates a consensus probability matrix using RNAfold [26]. Finally, the last k-nearest neighbor classifier provides a consensus score by utilizing the consensus probability matrix value and the base pair probabilities from previous classifiers. The secondary structure, corresponding to that score, is then accepted as a solution.

SVMs, originally developed by Vapnik [77], are a learning technique based on statistical learning theory and has its genesis emerged from the principle of percep-tron. SVMs, are used for mapping data points, that cannot be separated by a linear hyperplane, to a feature space so that the images of the data points in the feature space can be linearly separated. Considering a set of data points to be classified into two classes, a hyperplane is a generalization of the linear plane into a higher number of dimensions, such that, a good separation among data points is achieved by the hyperplane that has the largest distance to the training data points of any class. Most of the research works in predicting RNA secondary structure, using SVMs, are based on the alignment of RNA sequence with known sequence. In [92], RNA structure prediction problem is considered as a 2-class classification problem and SVMs are used to predict whether two columns of sequence alignment form a base pair or not. The hyperplane in SVM is constructed by training it with positive and negative sam-ples. The positive samples are those pair sites that form a base pair in the alignment of known sequences, and the negative samples are not. The feature vector for each pair site is composed of the covariation score, the base-pair fraction, and the base pair probability matrix. While, the covariation score is a measure of complementary mutations, considering evolutionary information in the two columns of an alignment, the fraction of complementary nucleotides show the bias toward base-pair for a pair of alignment columns. The base pair probability matrix is a complementarity for detecting the conserved base pairs and the base pair probability for every sequence in the alignment is computed with RNAfold [26]. These probability matrices are then aligned according to the sequence alignment and averaged. Considering the effect of sequence similarity upon covariation score, a similarity weight factor is also introduced, which adjusts the contribution of covariation and thermodynamic information toward prediction, based on sequence similarity. Finally, the common secondary structure is assembled by stem combing rules.

Table 7.1 Description of RNA sequences taken from the comparative RNA Web Site [7]

Organism	Accession number	RNA class	Length	Known bps
S. cerevisiae	X67579	5S rRNA	118	37
H. marismortui	AF034620	5S rRNA	122	38
M. anisopliae-2	AF197122	Group I intron, 23S rRNA	456	115
M. anisopliae-3	AF197120	Group I intron, 23S rRNA	394	120
A. lagunensis	U40258	Group I intron, 16S rRNA	468	113
H. rubra	L19345	Group I intron, 16S rRNA	543	138
A. griffini	U02540	Group I intron, 16S rRNA	556	131
C. elegans	X54252	16S rRNA	697	189
D. virilis	X05914	16S rRNA	784	233
X. laevis	M27605	16S rRNA	945	251
H. sapiens	J01415	16S rRNA	954	266
A. fulgens	Y08511	16S rRNA	964	265

7.8   Comparison Between Different Methods

Here, we compare the relative performance of some methods in predicting structures of 12 RNA sequences. These sequences are used in [73, 85], and they represent 12 different organisms, different sequence lengths and four different classes of RNA. The sequences are available in Comparative RNA web site [7] and details are provided in Table 7.1. The quality of a predicted RNA structure, from a given sequence, can be judged either by the number of accurately predicted base pairs or by the level of the minimized free energy, and the number of true positive (TP) base pairs and sensitivity can used as the RNA structure evaluation criterion and the performance evaluation criterion of the methods. While, the value of true positive (TP) base pairs, for a given RNA sequence, is the number of correctly identified base pairs among all predicted pairs, false positives are those non base pairs which are incorrectly identified as base pairs. True negatives are those non base pairs correctly identified as non base pairs and false negatives are those base pairs which are incorrectly identified as non base pairs. The sensitivity is defined as

$$sensitivity = \frac{true\ positives}{true\ positives + false\ negatives}. \tag{7.12}$$

Considering the availability of results on the same RNA sequences and ease of implementation, a comparison among genetic algorithm (GA), simulated annealing

(SA), Hopfield neural network (HNN) and mfold [95], in terms of predicted base pairs (predicted bps), true positive base pairs (TP) and sensitivity, is provided in Table 7.2. The performance results for GA, SA and mfold are taken from [73, 85]. The HNN is implemented in a similar way as it is mentioned in [39] and the results are reported. From the results in Table 7.2, it is concluded in [59] that the average results for true positive base pairs and sensitivity are comparable for GA, SA and mfold and their performances are superior to HNN. It is also found that SA performs a little better than GA and mfold when the number of known base pairs exceeds 180. The mfold performs the best when the number of known base pairs is less than 150.

Figures 7.2 and 7.3 compares the relative performances of genetic algorithms, simulated annealing, Hopfield neural network and mfold in terms of sensitivity versus the number of known base pairs and TP versus the number of known base pairs, respectively, for 12 different RNA sequences. From the figures it is found that the curves for simulated annealing, genetic algorithms, and mfold are comparable and are at the top of the figures, whereas, the curve for Hopfield neural network lies below all other curves.

7.9 Challenging Issues and Granular Networks

The three dimensional structure of RNA is a complex one and it presents a new set of computational challenges to the bioinformatics community. The ability to relate the structural properties with the functional properties [58] may provide the key in solving these challenges. The different approaches for predicting RNA structure, discussed so far, involve only the efforts of some soft computing tools in their individual capacity. One of the major challenges in soft computing, namely, the symbiotic integration of its components, is still not yet addressed in the existing RNA literature. In this regard, integrated tools like neuro-fuzzy, rough evolutionary network, rough-fuzzy evolutionary network, and rough fuzzy computing [55], may provide new directions in increasing the tractability in the application domain. It may also be mentioned here that, fuzzy set theoretic models try to mimic human reasoning and can provide decisions having close resemblance with that of human.

As mentioned in previous chapters, granular computing (GrC) has been proven to be a useful paradigm in mining data set, large in both size and dimension. When a problem involves incomplete, uncertain, and vague information, it may be difficult to differentiate distinct elements and one may find it convenient to consider the data as granules, representing a group of data points that have very similar characteristics, and performing operations on them [18]. These characteristics can be obtained from similarity, equality and closeness between the data points. For example, in neural networks, the self-organizing map (SOM) is a clustering technique that organizes the data in groups according to the underlying pattern. Each such group can be represented as an information granule. Although GrC is still not explored for RNA structure prediction, incorporation of GrC in structure prediction task may provide a conceptual framework for feature selection, classification and clustering of the RNA

Table 7.2 Comparison between soft computing methods for different RNA secondary structure using Individual Nearest Neighbor with Hydrogen Bonds (INN-HB) model

Sequence	Known bps	Predicted bps				TP				Sensitivity (%)			
		SA	GA	HNN	mfold	SA	GA	HNN	mfold	SA	GA	HNN	mfold
S. cerevisiae	37	39	39	38	41	33	33	19	33	89.2	89.2	50	89.2
H. marismortui	38	30	30	41	34	27	27	14	29	71.1	71.1	38	76.3
M. anisopliae-2	115	131	135	126	133	57	55	32	52	49.6	47.8	28	45.2
M. anisopliae-3	120	121	121	129	116	75	75	43	92	62.5	62.5	36	76.7
A. lagunensis	113	132	131	123	133	73	68	36	74	64.6	60.2	32	65.5
H. rubra	138	162	161	150	167	79	79	43	83	57.2	57.2	31	60.1
A. griffini	131	168	161	141	174	87	81	45	95	66.4	61.8	34	72.5
C. elegans	189	205	202	222	217	49	55	32	40	25.9	29.1	17	21.2
D. virilis	233	239	242	276	252	80	65	37	82	34.3	27.9	16	35.2
X. laevis	251	253	240	251	245	112	93	55	113	44.6	37.1	22	45.0
H. sapiens	266	244	250	266	258	116	89	53	95	43.6	33.5	20	35.7
A. fulgens	265	252	242	265	241	93	82	48	74	35.1	30.9	18	27.9
Averages	158.0	164.7	162.8	169.1	167.6	73.4	66.8	38.1	71.8	53.7	50.7	28.5	54.2

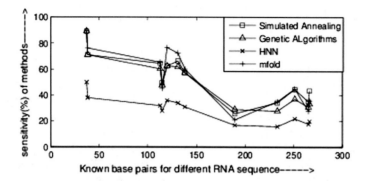

Fig. 7.2 Comparison among different methods in terms of sensitivity versus the number of known base pairs, for 12 different RNA sequences

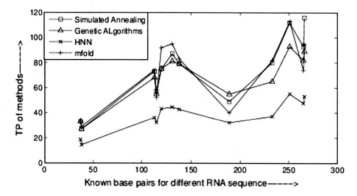

Fig. 7.3 Comparison among different methods in terms of true positive base pairs versus the number of known base pairs, for 12 different RNA sequences

sequence data. Here, fuzzy sets, rough sets and neural networks can be used in both, formulating granules and performing GrC.

Although the existing approaches for predicting RNA structure are useful, there is still some room for improving the output results. For example, in GAs, the basic crossover and mutation operators are common to all applications and can limit the effectiveness of GAs in structure prediction task; therefore, focused research to design more realistic and context sensitive operators is needed so that they can be coupled with the existing techniques. It should be mentioned here that, GAs and metahuristics are more suitable than HNN for global optimization based tasks. So, investigations in maximizing the base pairs for RNA structure, using GAs or SAs rather than HNN, may provide encouraging results. The future research of RNA informatics will require integration of different soft computing tools in an efficient manner to enhance the computational intelligence in solving the related prediction problems: thereby signifying the collaboration between soft computing and RNA communities.

7.10   Conclusion

Some existing methodologies in soft computing framework for RNA secondary structure prediction problem is presented from [59]. In this regard, the basic concepts in
RNA, different structural elements and the effect of ions, proteins and temperature on
the RNA molecule are discussed. Brief descriptions of some dynamic programming
based software packages are also provided. The relevance of certain soft computing tools, specially GAs, are more explored. The comparisons among some existing
methodologies, using 12 known RNA structures, revealed that average results are
comparable for GA, mfold and SA and they are superior to HNN. Future challenging
issues regarding the importance of relating the structural properties with the functional properties, integration of different soft computing methodologies, application
of GAs and different metahuristics in solving maximum independent set for prediction of RNA structure, and the need to design structure specific operators, are
addressed.

In some of the investigations, the hybridization of dynamic programming (DP)
with FL, GAs and metahuristics, revealed a new research direction [59]. First, substructures were predicted using DP and then the optimal or suboptimal structures
were estimated using soft computing techniques. GAs appear to be a powerful soft
computing tool to handle the task of structure prediction by not only considering the
kinetic effects and cotranscriptional folding, but also for estimation of certain free
energy parameters. The existing metahuristics deal with permutation of substructures
with minimum free energy, but they have the potential to explore RNA folding pathways by itself creating the stems, pseudostems and temporary stems, as performed
by various GAs. They can also be utilized for predicting the pseudoknots and energy
parameters. Although, individual methods can compute the optimal or suboptimal
structures within a given thermodynamic model, the natural fold of RNA is often in
an energy state related to cotranscriptional folding or kinetic energy traps in folding
landscape, and requires soft computing based methods to achieve those states.

References

1. Akutsu, T.: Dynamic programming algorithms for RNA secondary structure prediction with
   pseudoknots. Discrete Appl. Math. **104**, 45–62 (2000)
2. Bakin, A., Ofengand, J.: Four newly located pseudouridylate residues in Escherichia coli 23S
   ribosomal RNA are all at the peptidyltransferase center: analysis by the application of a new
   sequencing technique. Biochemistry **32**, 9754–62 (1993)
3. Batenburg, F.H.V., Gultyaev, A.P., Pleij, C.W.: An APL-programmed genetic algorithm for the
   prediction of RNA secondary structure. J. Theor. Biol. **174**(3), 269–280 (1995)
4. Batey, R.T., Gilbert, S.D., Montange, R.K.: Structure of a natural guanine-responsive riboswitch
   complexed with the metabolite hypoxanthine. Nature **432**, 411–415 (2004)
5. Bindewald, E., Shapiro, B.A.: RNA secondary structure prediction from sequence alignments
   using a network of k-nearest neighbor classifiers. RNA **12**, 342–352 (2006)
6. Booker, L.B., Goldberg, D.E., Holland, J.H.: Classifier systems and genetic algorithms. Artif.
   Intell. **40**, 235–282 (1989)

7. Cannone, J., Subramanian, S., Schnare, M., Collett, J., Dśouza, L., Du, Y., Feng, B., Lin, N., Madabusi, L., Muller, K., Pande, N., Shang, Z., Yu, N., Gutell, R.: The comparative RNA web (CRW) site: an online database of comparative sequence and structure information for ribosomal, intron, and other RNAS. BMC Bioinform. **3**(1), 2 (2002)

8. Crick, F.H.: Codon-anticodon pairing: the wobble hypothesis. J. Mol. Biol. **19**, 548–55 (1966)

9. Daou-Chabo, R., Condon, C.: RNase J1 endonuclease activity as a probe of RNA secondary structure. RNA **15**, 1417–25 (2009)

10. Ding, Y., Chan, C.Y., Lawrence, C.E.: Sfold web server for statistical folding and rational design of nucleic acids. Nucleic Acids Res. **32**, W135–W141 (2004)

11. Ding, Y., Chan, C.Y., Lawrence, C.E.: RNA secondary structure prediction by centroids in a Boltzmann weighted ensemble. RNA **11**(8), 1157–1166 (2005)

12. Ding, Y., Lawrence, C.E.: A statistical sampling algorithm for RNA secondary structure prediction. Nucleic Acids Res. **31**, 7280–7301 (2003)

13. Doetsch, M., Schroeder, R., Furtig, B.: Transient RNA protein interactions in RNA folding. FEBS J. **278**(4), 1634–1642 (2011)

14. Doherty, E.A., Batey, R.T., Masquida, B., Doudna, J.A.: A universal mode of helix packing in RNA. Nat. Struct. Biol. **8**, 339–343 (2001)

15. Doshi, K.J., Cannone, J.J., Cobaugh, C.W., Gutell, R.R.: Evaluation of the suitability of free-energy minimization using nearest-neighbor energy parameters for RNA secondary structure prediction. BMC Bioinform. **5**(105), 1–22 (2004)

16. Esogbue, A.O., Kacprzyk, J.: Fuzzy dynamic programming: main developments and applications. Fuzzy Sets Syst. **81**, 31–45 (1996)

17. Ferentz, A.E., Wagner, G.: Nmr spectroscopy: a multifaceted approach to macromolecular structure. Q. Rev. Biophys. **33**, 29–65 (2000)

18. Ganivada, A., Ray, S.S., Pal, S.K.: Fuzzy rough granular self-organizing map and fuzzy rough entropy. Theoret. Comput. Sci. **466**, 37–63 (2012)

19. Goldberg, D.: Genetic Algorithms in Optimization, Search, and Machine Learning. Addison Wesley, Reading (1989)

20. Grner, W., Giegerich, R., Strothmann, D., Reidys, C., Weber, J., Hofacker, I.L., Stadler, P.F., Schuster, P.: Analysis of RNA sequence structure maps by exhaustive enumeration. II. structures of neutral networks and shape space covering. Monatsh. Chem. Mon. **127**, 375–389 (1996). doi:10.1093/nar/gks468

21. Grner, W., Giegerich, R., Strothmann, D., Reidys, C., Weber, J., Hofacker, I.L., Stadler, P.F., Schuster, P.: Modeling and automation of sequencing-based characterization of RNA structure. Proc. Natl. Aca. Sci. USA **108**(27), 11,069–11,074 (2011)

22. Gultyaev, A.P., Batenburg, F.H.V., Pleij, C.W.: The computer simulation of RNA folding pathways using an genetic algorithm. J. Mol. Biol. **250**, 37–51 (1995)

23. Haykin, S.: Neural Networks: A Comprehensive Foundation, 2nd edn. Prentice Hall, Upper Saddle River (1998)

24. Haynes, T., Knisley, D., Knisley, J.: Using a neural network to identify secondary RNA structures quantified by graphical invariants. MATCH Commun. Math. Comput. Chem. **60**, 277–290 (2008)

25. Hofacker, I.L., Fontana, W., Stadler, P.F., Bonhoeffer, L.S., Tacker, M., Schuster, P.: Fast folding and comparison of RNA secondary structures. Monats. Chem. Mon. **125**, 167–188 (1994)

26. Hofacker, I.L.: Vienna RNA secondary structure server. Nucleic Acids Res. **31**, 3429–31 (2003)

27. Holley, R.W., Apgar, J., Everett, G.A., Madison, J.T., Marquisee, M., Merrill, S.H., Penswick, J.R., Zamir, A.: Structure of ribonucleic acid. Science **147**, 1462–1465 (1965)

28. Huber, P.W.: Chemical nucleases: their use in studying RNA structure and RNA-protein interactions. FASEB J. **7**, 1367–1375 (1993)

29. Karaduman, R., Fabrizio, P., Hartmuth, K., Urlaub, H., Lührmann, R.: RNA structure and RNA-protein interactions in purified yeast U6 snRNPs. J. Mol. Biol. **356**, 1248–62 (2006)

30. Kasprzak, W., Shapiro, B.A.: Stem Trace: an interactive visual tool for comparative RNA structure analysis. Bioinformatics **15**, 16–31 (1999)

31. Kim, S.H., Quigley, G., Suddath, F.L., Rich, A.: High-resolution X-Ray diffraction patterns of crystalline transfer RNA that show helical regions. Proc. Natl. Aca. Sci. USA **68**, 841–845 (1971)
32. Kirkpatrick, S., Gelatt, C.D., Vecchi, M.P.: Optimization by simulated annealing. Science **220**(4598), 671–680 (1983)
33. Koculi, E., Cho, S.S., Desai, R., Thirumalai, D., Woodson, S.A.: Folding path of P5abc RNA involves direct coupling of secondary and tertiary structures. Nucleic Acids Res. 1–10 (2012). doi:10.1093/nar/gks468
34. Koessler, D.R., Knisley, D.J., Knisley, J., Haynes, T.: A predictive model for secondary RNA structure using graph theory and a neural network. BMC Bioinform. **11**, S6–S21 (2010)
35. Le, S., Nussinov, R., Maziel, J.: Tree graphs of RNA secondary structures and their comparison. Comput. Biomed. Res. **22**, 461–473 (1989)
36. Lehninger, A.L., Nelson, D.L., Cox, M.M.: Principles of Biochemistry. Worth, New York (1993)
37. Leontis, N.B., Westhof, E.: A common motif organizes the structure of multi-helix loops in 16S and 23S ribosomal RNAs. J. Mol. Biol. **283**, 571–583 (1998)
38. Leontis, N.B., Westhof, E.: Geometric nomenclature and classification of RNA base pairs. RNA **7**, 499–512 (2001)
39. Liu, Q., Ye, X., Zhang, Y.: A Hopfield neural network based algorithm for RNA secondary structure prediction. In: Proceedings of the 1st International Conference on Multi-Symposiums on Computer and Computational Sciences, pp. 1–7 (2006)
40. Lodish, H., Berk, A., Kaiser, C.A., Krieger, M., Scott, M.P., Bretscher, A., Ploegh, H., Matsudaira, P.: Molecular Cell Biology, 5th edn. W. H. Freeman, New York (2003)
41. Low, J.T., Weeks, K.M.: Shape-directed RNA secondary structure prediction. Methods **52**(2), 150–158 (2010)
42. Mathews, D.H., Sabina, J., Zuker, M., Turner, D.H.: Expanded sequence dependence of thermodynamic parameters improves prediction of rna secondary structure. J. Mol. Biol. **288**, 911–940 (1999)
43. Mathews, D.H., Disney, M.D., Childs, J.L., Schroeder, S.J., Zuker, M., Turner, D.H.: Incorporating chemical modification constraints into a dynamic programming algorithm for prediction of RNA secondary structure. Proc. Natl. Acad. Sci. USA **101**, 7287–7292 (2004)
44. Mathews, D.H.: Predicting RNA secondary structure by free energy minimization. Theor. Chem. Acc. Theor. Comput. Model. **116**, 160–168 (2006)
45. McCaskill, J.S.: The equilibrium partition function and base pair binding probabilities for RNA secondary structure. Biopolymers **29**, 1105–1119 (1990)
46. Meyer, I.M., Miklos, I.: Co-transcriptional folding is encoded within RNA genes. BMC Mol. Biol. **5**, 1–10 (2004). doi:10.1186/1471-2199-5-10
47. Mitchell, M., Forrest, S., Holland, J.H.: The royal road for genetic algorithms: fitness landscapes and GA performance. In: Proceedings of the 1st European Conference on Artificial Life (1992)
48. Needleman, S.B.: A general method applicable to the search for similarities in the amino acid sequence of two proteins. J. Mol. Biol. **48**(3), 443–53 (1970)
49. Neethling, M., Engelbrecht, A.: Determining RNA secondary structure using set-based particle swarm optimization. In: IEEE Congress on Evolutionary Computation, pp. 6134–6141 (2006)
50. Noller, H.F., Chaires, J.B.: Functional modification of 16s ribosomal RNA by kethoxal. Proc. Natl. Acad. Sci. USA **69**, 3115–3118 (1972)
51. Nussinov, R., Pieeznik, G., Griggs, J.R., Kleitman, D.J.: Algorithms for loop matching. SIAM J. Appl. Math. **35**, 68–82 (1978)
52. Nussinov, R., Jacobson, A.B.: Fast algorithm for predicting the secondary structure of single-stranded RNA. Proc. Natl. Aca. Sci. USA **77**(11), 6309–6313 (1980)
53. Pal, S.K., Bandyopadhyay, S., Ray, S.S.: Evolutionary computation in bioinformatics: a review. IEEE Trans. Syst. Man Cybern. Part C **36**(5), 601–615 (2006)
54. Pal, S.K., Ghosh, A.: Soft computing data mining. Inf. Sci. **163**, 1–3 (2004)
55. Pal, S.K., Skowron, A.: Rough-Fuzzy Hybridization: A New Trend in Decision Making. Springer, New York (1999)

56. Rambo, R.P., Tainer, J.A.: Improving small-angle X-ray scattering data for structural analyses of the RNA world. RNA **16**, 638–46 (2010)
57. Ray, S.S., Bandyopadhyay, S., Mitra, P., Pal, S.K.: Bioinformatics in neurocomputing framework. IEE Proc. Circuits Devices Syst. **152**, 556–564 (2005)
58. Ray, S.S., Halder, S., Kaypee, S., Bhattacharyya, D.: HD-RNAS: an automated hierarchical database of RNA structures. Front. Genet. **3**(59), 1–10 (2012)
59. Ray, S.S., Pal, S.K.: RNA secondary structure prediction using soft computing. IEEE/ACM Trans. Comput. Biol. Bioinf. **10**(1), 2–17 (2013)
60. Regulski, E.E., Breaker, R.R.: In-line probing analysis of riboswitches. Meth. Mol. Biol. **419**, 53–67 (2008)
61. Schmitz, M., Steger, G.: Description of RNA folding by simulated annealing. J. Mol. Biol. **255**(1), 254–66 (1996)
62. Schultes, E.A., Bartel, D.P.: One sequence, two ribozymes: implications for the emergence of new ribozyme folds. Science **289**, 448–452 (2000)
63. Shapiro, B.A., Bengali, D., Kasprzak, W., Wu, J.C.: RNA folding pathway functional intermediates: their prediction and analysis. J. Mol. Biol. **312**, 27–44 (2001)
64. Shapiro, B.A., Wu, J.C., Bengali, D., Potts, M.J.: The massively parallel genetic algorithm for RNA folding: MIMD implementation and population variation. Bioinformatics **17**(2), 137–148 (2001)
65. Shapiro, B.A., Navetta, J.: A massively parallel genetic algorithm for RNA secondary structure prediction. J. Supercomput. **8**, 195–207 (1994)
66. Shapiro, B.A., Wu, J.C.: An annealing mutation operator in the genetic algorithms for RNA folding. J. Supercomput. **12**, 171–180 (1996)
67. Shapiro, B.A., Wu, J.C.: Predicting h-type pseudoknots with the massively parallel genetic algorithm. Comput. Appl. Biosci. **13**, 459–471 (1997)
68. Singer, B.: All oxygens in nucleic acids react with carcinogenic ethylating agents. Nature **264**, 333–339 (1976)
69. Song, D., Deng, Z.: A fuzzy dynamic programming approach to predict RNA secondary structure. In: Algorithms in Bioinformatics: Lecture Notes in Computer Science, vol. 4175, pp. 242–251 (2006)
70. Tijerina, P., Mohr, S., Russell, R.: DMS footprinting of structured RNAs and RNA-protein complexes. Nat. Protoc. **2**, 2608–23 (2007)
71. Tinoco, I.J., Borer, P.N., Dengler, B., Levin, M.D., Uhlenbeck, O.C., Crothers, D.M., Bralla, J.: Improved estimation of secondary structure in ribonucleic acids. Nat. New Biol. **246**, 40–41 (1973)
72. Tinoco, T., Bustamante, C.: How RNA folds. J. Mol. Biol. **293**(1), 271–281 (1999)
73. Tsang, H.H., Wiese, K.C.: SARNA-predict: accuracy improvement of RNA secondary structure prediction using permutation based simulated annealing. IEEE/ACM Trans. Comput. Biol. Bioinf. **7**(4), 727–740 (2010)
74. Tullius, T.D., Dombroski, B.A.: Hydroxyl radical footprinting: high-resolution information about DNA-protein contacts and application to lambda repressor and Cro protein. Proc. Natl. Acad. Sci. USA **83**, 5469–5473 (1986)
75. Turner, D.H., Mathews, D.H.: NNDB: the nearest neighbor parameter database for predicting stability of nucleic acid secondary structure. Nucleic Acids Res. **38**, D280–D282 (2009)
76. Turner, D.H., Sugimoto, N.: RNA structure prediction. Annu. Rev. Biophys. Biophys. Chem. **17**, 167–192 (1988)
77. Vapnik, V.: The Nature of Statistical Learning Theory. Springer, New York (1995)
78. Varani, G., McClain, W.H.: The G x U wobble base pair. A fundamental building block of RNA structure crucial to RNA function in diverse biological systems. EMBO Rep. **1**, 18–23 (2000)
79. Waterman, M.S.: RNA secondary structure: a complete mathematical analysis. Math. Biosci. **42**, 257–266 (1978)
80. Waterman, M., Smith, T.: Rapid dynamic programming algorithms for RNA secondary structure. Adv. Appl. Math. **7**(0196–8858/86), 455–464 (1986)

81. Watson, J.D., Crick, F.H.C.: Molecular structure of nucleic acids: a structure for deoxyribose nucleic acid. Nature **171**, 737–738 (1953)
82. Westhof, E., Auffinger, P.: RNA tertiary structure, encyclopedia of analytical chemistry. In: Meyers, R.A. (Ed.), pp. 5222–5232. Wiley, Chichester (2000)
83. Westhof, E., Masquida, B., Jossinet, F.: Predicting and modeling RNA architecture. Cold Spring Harb. Perspect. Biol. **a003632**, 1–12 (2011)
84. Wiese, K.C., Deschenes, A., Glen, E.: Permutation based RNA secondary structure prediction via a genetic algorithm. In: Proceedings of the 2003 Congress on Evolutionary Computation, pp. 335–342 (2003)
85. Wiese, K.C., Deschênes, A.A., Hendriks, A.G.: RnaPredict-an evolutionary algorithm for RNA secondary structure prediction. IEEE/ACM Trans. Comput. Biol. Bioinf. **5**(1), 25–41 (2008)
86. Wiese, K.C., Glen, E.: A permutation-based genetic algorithm for the RNA folding problem: a critical look at selection strategies, crossover operators, and representation issues. Biosystems **72**, 29–41 (2003)
87. Wiese, K.C., Hendriks, A.: Comparison of P-RnaPredict and mfold-algorithms for RNA secondary structure prediction. Bioinformatics **22**(8), 934–942 (2006)
88. Xayaphoummine, A., Bucher, T., Isambert, H.: Kinefold web server for RNA/DNA folding path and structure prediction including pseudoknots and knots. Nucleic Acids Res. **33**, W605–W610 (2005). doi:10.1186/1471-2199-5-10
89. Zadeh, L.A.: Fuzzy sets. Inf. Control **8**, 338–353 (1965)
90. Zadeh, L.A.: Fuzzy logic, neural networks, and soft computing. Commun. ACM **37**, 77–84 (1994)
91. Zhang, G.P.: Neural networks for classification: a survey. IEEE Trans. Syst. Man Cybernet. Part C **30**(4), 451–462 (2000)
92. Zhao, Y., Wang, Z.: Consensus RNA secondary structure prediction based on support vector machine classification. Chin. J. Biotechnol. **24**(7), 1140–1148 (2008)
93. Zou, Q., Zhao, T., Liu, Y., Guo, M.: Predicting RNA secondary structure based on the class information and Hopfield network. Comput. Biol. Med. **39**(3), 206–214 (2009)
94. Zuker, M.: On finding all suboptimal foldings of an rnamolecule. Science **244**, 48–52 (1989)
95. Zuker, M.: Mfold web server for nucleic acid folding and hybridization prediction. Nucleic Acids Res. **31**, 3406–3415 (2003)
96. Zuker, M., Stiegler, P.: Optimal computer folding of large RNA sequences using thermodynamics and auxiliary information. Nucleic Acids Res. **9**(1), 133–148 (1981)

Appendix

Various data sets used in the investigation are briefly described below.

Telugu vowel data: The speech data deals with 871 Indian Telugu vowel sounds. These were uttered in a consonant-vowel-consonant context by three male speakers in the age group of 30–35 years. The dataset has three features, say F_1, F_2 and F_3, corresponding to the first, second and third vowel format frequencies obtained through spectrum analysis of speech data. There are six vowel classes, denoted by δ, a, i, u, e and o.

Gene expression microarray data sets: For microarray gene expression analysis, Cell Cycle, Yeast Complex and All Yeast data sets are chosen. The genes in these data sets belong to Saccharomyces cerevisiae and are classified into 16, 16 and 18 groups, respectively, according to the functional annotations of Munich Information for Protein Sequences (MIPS) database.

Sonar: Bouncing sonar signals of a metal cylinder at various angles and under various conditions represent patterns. The frequencies of sonar signals transmitted to metal cylinder are attribute values ranging from 0 to 1. There are two classes of signals obtained from two different angles, spanning 90° for the cylinder and 180° for the sonar rock.

Lymphography: The structures of lymphatic system including lymph nodes, lymph ducts, and lymph vessels are visualized in an X-ray of an injected radio contrast image. The image of the vessels and nodes is called lymphogram. Four classes, normal find, metastases, malign lymph and fibrosis, of patterns with eighteen attributes (different values of vessels and nodes) are available with this data.

Glass: This data contains of patterns made of float processed windows and not float processed windows. There are nine attributes with every pattern.

Thyroid: This data contains an analysis of the thyroid decease patients. There are 215 patients with five attributes belonging to three categories.

Spam base: This data contains 4601 instances belonging to spam and non spam emails. Each instant has 57 continuous valued attributes representing word frequencies.

© Springer International Publishing AG 2017
S.K. Pal et al., *Granular Neural Networks, Pattern Recognition
and Bioinformatics*, Studies in Computational Intelligence 712,
DOI 10.1007/978-3-319-57115-7

Iris: The iris data consists of 150 patterns with 4 features/attributes and 3 classes. Each class contains 50 patterns, referring to a type of iris plant, and is linearly separable from the other classes. Attribute information is given in terms of sepal length, sepal width, petal length and petal width representing the four features.

Wisconsin Cancer: This data consists of 684 patterns and 9 features. All patterns belong to 2 classes, represented by benign tumor and malignant tumor.

Waveform: This data contains 5000 patterns, 40 attributes and 3 wave classes. While, all the attribute values are continuous between 0 and 6, 19 attributes among them are noise values with mean 0 and variance 1.

Multiple Features: This data set has 2000 patterns with 625 features of handwritten numerals, ('0'–'9'), extracted from a collection of Dutch utility maps and 10 classes. The values of attributes are integers and real type.

ALL and AML data: The data involves 27 acute myelogenous leukaemia (AML) and 11 acute lymphoblastic leukaemia (ALL) samples. Each sample is generated with 5000 gene expression values which shows the differences between AML and ALL samples.

Prostate Cancer: The data contains 52 tumor and 50 normal samples. The expression profiles for 12,600 genes are derived from both the tumor and normal samples using the oligonucleotide microarrays.

Lung cancer: This data contains total 197 samples, out of which 139 adenocarcinoma, 21 squamous cell carcinoma, 20 carcinoid and 17 normal lung samples. Each sample has 1000 gene expression values. The gene expressions profiles of the normal and cancer sample are based on oligonucleotide array.

Multiple Tissue Type (Multi-A): The multiple tissue type-A (Multi-A) data contains 103 samples belonging to four different tissue types like, breast, prostate, lung, and colon, representing four classes or groups. Here, 5565 gene expressions are available for each sample.

It may be noted that, the last 4 data sets (from ALL and AML to Multiple Tissue Type) are downloaded from the following links http://www.broadinstitute.org/cgi-bin/cancer/datasets.cgi/, http://www.ncbi.nlm.nih.gov/sites/GDSbrowser, and http://genomics-pubs.princeton.edu/oncology/, respectively.

Index

© Springer International Publishing AG 2017
S.K. Pal et al., *Granular Neural Networks, Pattern Recognition
and Bioinformatics*, Studies in Computational Intelligence 712,
DOI 10.1007/978-3-319-57115-7

Printed in the United States
By Bookmasters